ORDER NUMBER EA-BEM

BASIC ELECTRONICS
AND RADIO INSTALLATION

International Standard Book Number 0-89100-064-X
For sale by: IAP, Inc.
Mail to: P.O. Box 10000, Casper, WY 82602-1000
Ship to: 7383 6WN Road, Casper, WY 82604-1835
(800) 443-9250 • (307) 266-3838 • FAX: 307-472-5106

HBC0393 Printed in the USA

Library of Congress Cataloging-in-Publication Data

Basic electronics and radio installation.
     p. cm. -- (Aviation technician training series.)   (An IAP,
Inc. training manual)
  ISBN 0-89100-064-X
   1. Avionics--Programmed instruction.   2. Radio in aeronautics-
-Programmed instruction.   I. Series.   II. Series: An IAP, Inc.
training manual.
TL695.B344 1992
629.135'1--dc20   92-24628
                                             CIP

Table of Contents

Preface vi

Introduction vii

   I.   Review of Alternating Current Electricity 1-1

       A. Inductance 1-1

       B. Capacitance 1-1

       C. Impedance 1-2

       D. Resonance 1-2

  II. Electron Control Valves 2-1

       A. Vacuum tubes 2-1

           1. Diodes 2-1

           2. Triodes 2-1

           3. Tetrodes 2-2

           4. Pentodes 2-2

       B. Semiconductor Diodes 2-3

       C. Zener Diodes 2-5

       D. Transistors 2-6

       E. Silicon-Controlled Rectifiers 2-7

       F. Triacs 2-8

       G. Field Effect Transistors 2-8

       H. Tunnel Diodes 2-9

       I. Unijunction Transistors 2-9

       J. Optioelectronic Devices 2-10

           1. Photodiodes 2-10

           2. Phototransistors 2-10

           3. Light emitting diodes 2-11

III. Practical Circuits 3-1

    A. Rectifier Circuits 3-1

        1. Half-wave rectifiers 3-1

        2. Full-wave rectifiers 3-1

            a. Two-diode 3-1

            b. Four-diode, bridge type 3-1

            c. Three-phase 3-2

    B. Amplifier Circuits 3-2

        1. Common emitter amplifier 3-2

        2. Common base amplifier 3-3

        3. Common collector amplifier 3-3

    C. Oscillator Circuits 3-4

        1. Wave forms 3-4

            a. Sine wave 3-4

            b. Square wave 3-5

            c. Sawtooth wave 3-5

        2. Electronic oscillation 3-5

            a. Hartley oscillator 3-5

                (1) Series 3-5

                (2) Shunt 3-6

                (3) Crystal 3-7

            b. Multivibrator oscillator 3-7

                (1) Astable 3-7

                (2) Bistable 3-8

                (3) Monostable 3-8

    D. Filter Circuits 3-9

    E. Voltage Multipliers 3-10

        1. Half-wave voltage doubler 3-10

        2. Full-wave voltage doubler 3-11

F. Voltage Regulators 3-11

IV. Logic Circuits 4-1

A. Binary Number System 4-1

B. Logic Gates 4-2

1. AND gate 4-2

2. OR gate 4-2

3. Amplifier 4-2

4. NOT gate 4-3

5. NAND gate 4-3

6. NOR gate 4-3

7. EXCLUSIVE OR gate 4-3

V. Radio Communications 5-1

A. Radio Waves 5-1

1. Composition 5-2

a. Carrier waves 5-2

b. Modulation 5-3

2. Propagation and reception 5-4

B. Radio Receivers 5-4

1. Amplitude Modulation 5-4

a. Principle 5-4

b. Superheterodyne Receivers 5-5

c. Double Conversion Superheterodyne Receivers ...5-6

2. Frequency Modulation 5-7

a. Principle 5-7

b. FM Receivers 5-7

c. FM Uses 5-8

C. Radio Transmitters 5-8

D. Antenna 5-9

    1. Principle 5-9

    2. Length 5-9

    3. Polarization and Field Pattern 5-10

    4. Types 5-10

        a. Hertz 5-10

        b. Marconi 5-10

        c. Loop 5-11

    5. Voltage Standing Wave Ratio (VSWR) 5-12

E. Transmission Lines 5-13

VI. Radio Navigation 6-1

A. Very High Frequency Omnirange
   Radio Navigation (VOR) 6-1

B. Automatic Direction Finder (ADF) 6-3

C. Instrument Landing System (ILS) 6-5

    1. Localizer 6-5

    2. Compass Locaters 6-6

    3. Marker Beacons 6-7

    4. Glide Slope 6-8

D. Distance Measuring Equipment 6-9

E. Radar Beacon Transponder 6-10

VII. Electronics Installation 7-1

A. Electrical Considerations 7-1

    1. Load Analysis 7-1

    2. Circuit Protection 7-1

    3. Wiring 7-2

    4. Wire Termination 7-3

        a. Terminal strips 7-3

        b. Connector plugs 7-4

     5.   Shielding 7-5

     6.   Bonding 7-5

     7.   Bundling and routing 7-6

  B. Mounting 7-7

  C. Antenna Installation 7-8

  D. Electrical and Magnetic Interference 7-8

  E. Paperwork 7-9

Glossary A-1

Answers to Study Questions B-1

Final Examination C-1

Answers to Final Examination D-1

Appendix: Electronic Symbols E-1

Preface

This book on *Basic Electronics and Radio Installation* is one of a series of specialized training manuals prepared for aviation maintenance personnel.

This series is part of a programmed learning course developed and produced by IAP, Inc., to improve and promote the aviation maintenance industry through research, communications, and education. This program is part of the effort to improve the quality of education for aviation technicians throughout the world.

The purpose of each IAP training series is to provide basic information on the operation and principles of the various aircraft systems and their components. Texts such as this one are general in nature and provide background information in which the A&P can build his understanding of aircraft.

Throughout this text, at appropriate points, is included a series of carefully prepared questions and answers to emphasize key elements of the text, and to encourage the individual to continually test himself for accuracy and retention as he progresses. A multiple choice, final examination is included to allow you to test your comprehension of the total material.

If you have any questions or comments regarding this manual, or any of the many other textbooks offered by IAP, simply contact: Sales Department, IAP, Inc.; Mailing Address: P.O. Box 10000, Casper, WY 82602-1000; Shipping Address: 7383 6WN Road, Casper, WY 82604-1835; or call toll free: (800) 443-9250; International, call: (307) 266-3838.

Introduction

As aviation maintenance came of age, the A&P mechanic still had his hands full with mechanical problems, and he stayed away from the electrical systems as much as possible. But as the utility of the airplane increased, so did the amount of electrical equipment. The progressive A&P has therefore, had to change his attitude and learn everything possible about this invisible power.

Electronics, and elaboration of electricity, has become one of the vital aspects of aviation, and we can no more ignore it than we could ignore electrical systems a generation ago.

The airlines have their Electronic-Radio-Instrument (E-R-I) specialists, and the military services also have specialists available. But in general aviation, the A&P technician must service many electronically controlled systems, and it is important that he understand the background of this intriguing science.

Electricity, as we mechanics think of it, is a flow of electrons along conductors, controlled by switches, fuses, circuit breakers or resistors, and whose characteristics may be modified by capacitors or coils which shift the phase between the pressure and the flow. When we think of electronics, we still consider the flow of electrons, but this time we replace some of the mechanical components with tubes, solid state diodes, transistors, or integrated circuit chips.

The material covered in this manual assumes a basic knowledge of electricity. It uses only the mathematics required for an understanding of the *principles,* and as much as possible using mechanical analogies to explain electronic functions.

Be sure to check your progress with the questions scattered throughout the text, as electronics, even more than many other subjects, builds on material previously mastered.

SECTION I:

Review of Alternating Current Electricity

A. INDUCTANCE

When current flows in a conductor, magnetic fields, or lines of flux, radiate out and encircle it. The energy in these fields generates a voltage in any conductor they move across, and this voltage, called counter EMF, is opposite in polarity to the voltage causing the original current. Its value is determined by the rate at which the lines of flux cross the conductor. If the conductor is wound into a coil and some material having a high permeability is used as a core, the lines of flux will be concentrated and the voltage will be increased.

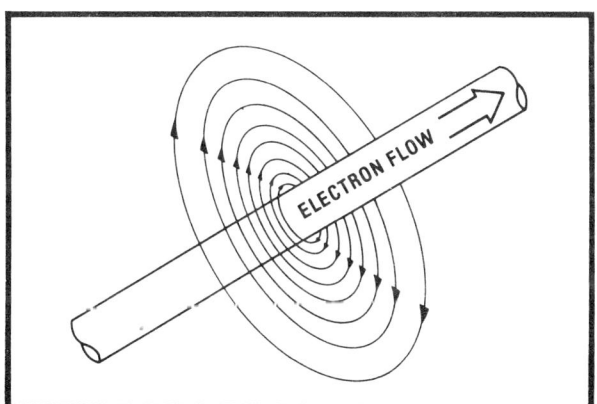

Fig. 1 *A magnetic field surrounds a conductor carrying current. The field is perpendicular to the conductor, and its strength is proportional to the amount of current.*

Counter EMF is generated only when current is *changing* and the lines of flux are moving in or out. When alternating current is applied to an inductor (coil), the current continually changes and the fastest *rate of change*, therefore the greatest voltage, occurs at the time the current passes through zero, Fig. 2. The voltage and current in an inductor will be 90 degrees out of phase with each other, and the voltage change will occur ahead of the current change.

Inductance causes an opposition to flow of current in an AC circuit, and this opposition, called inductive reactance, is a function of both the inductance and the frequency. The formula for

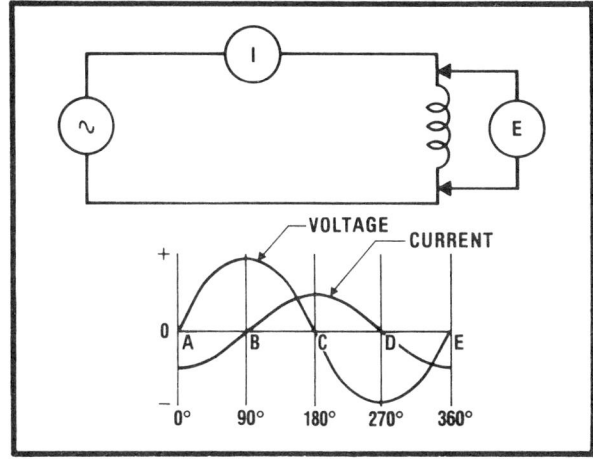

Fig. 2 *In a purely inductive circuit, the voltage change leads the current change by 90°.*

inductive reactance is $X_L = 2\pi FL$, which inductive reactance is expressed in ohms, frequency in hertz, and inductance in henries. Two pi is a constant, 6.28.

B. CAPACITANCE

When two conductors are separated by an insulator, a capacitor is formed. A capacitor, or condenser, as it was formerly called, stores electrical energy in the form of electrostatic fields.

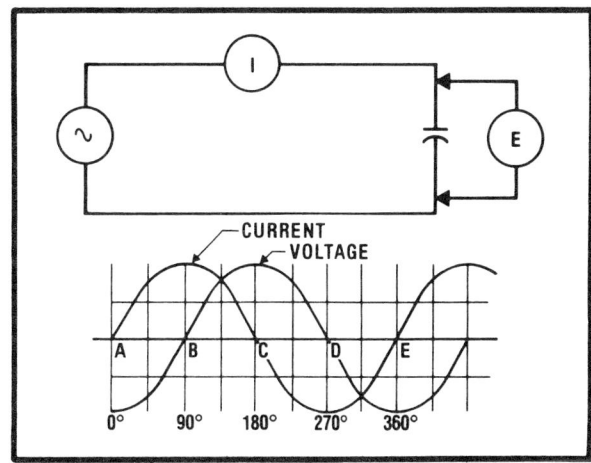

Fig. 3 *In a purely capacitive circuit, the current change leads the voltage change by 90°.*

The amount it stores is determined by the area of the plates (the conductors), the distance between the plates, and the material that separates them. As electrons flow into a capacitor to charge it, the voltage across it rises, and, as you notice in Fig. 3, in a purely capacitive circuit the current changes lead the voltage changes by 90°. Capacitance opposes the flow of AC because the current flows into and out of the capacitor instead of in the rest of the circuit. The larger the capacitor and the lower the frequency, the greater the opposition.

Capacitive reactance: $$X_C = \frac{1}{2\pi FC}$$

In this formula, as in that for inductive reactance, the reactance, or opposition, is expressed in ohms and the frequency in hertz; but here, the capacitance is expressed in farads.

C. IMPEDANCE

For all practical purposes, no circuit can contain only capacitance or inductance. All circuits contain some resistance, in which the voltage and current are in phase; some inductance, with the voltage leading the current; and some capacitance, with the current leading the voltage. The result of all three of these oppositions is equal to their vector sum, and this is called impedance, Z. Since both inductive and capacitive reactance shift the phase by 90°, and these shifts are in the opposite direction, the two reactances are 180° apart and tend to cancel each other.

The impedance of a circuit is the square root of the sum of the resistance squared, plus the reactance squared.

$$Z = \sqrt{(X_L - X_C)^2 + R^2} \quad \text{or} \quad Z = \sqrt{(X_C - X_L)^2 + R^2}$$

D. RESONANCE

As the frequency of alternating current increases, the inductive reactance becomes greater because the lines of flux expand and contract faster,

producing a greater *rate* of flux change. The capacitive reactance, however, will become less because the capacitor can charge up less completely. There is a frequency at which the two reactances are the same, and this is called the resonant frequency. To find the resonant frequency of a circuit, use the formula:

$$Fr = \frac{1}{2\pi\sqrt{LC}}$$

Fr = Resonant frequency in hertz
2π = A constant, 6.28
L = Inductance in henries
C = Capacitance in farads

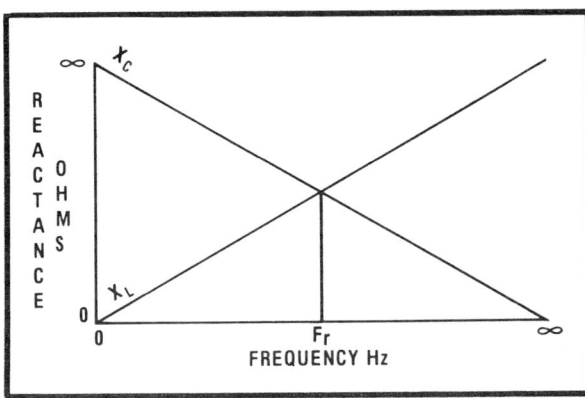

Fig. 4 *Resonant frequency is that frequency at which the inductive and capacitive reactances are the same.*

QUESTIONS:

1. In an inductive circuit, do the current changes lead or lag the voltage changes causing them?

2. Does capacitive reactance become larger or smaller as the frequency increases?

3. What is the name of the vector sum of reactance and resistance?

4. What is the name of the frequency at which inductive reactance and capacitive reactance are equal?

SECTION II:

Electron Control Valves

A. VACUUM TUBES

Vacuum tubes have just about passed from the scene as far as aircraft electronics is concerned, but because of the principles involved, we will take a quick look at them.

1. Diodes

The filament, or heater, heated red hot with low-voltage electricity, has no function other than to heat the cathode, the small tube surrounding the heater. The surface of the cathode is covered with a material whose electrons are easily excited by heat, and when heated, expand their orbits so much that they can be drawn away by a positive voltage on a nearby electrode called the plate, Fig. 5.

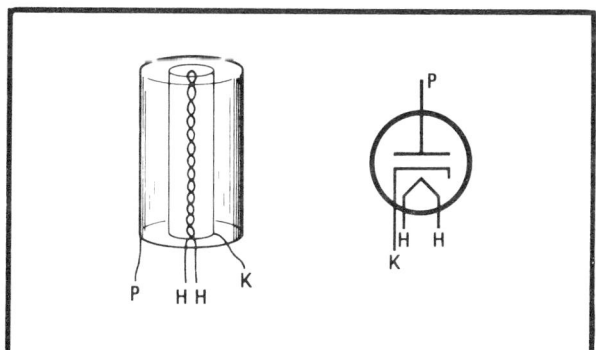

Fig. 5 In a diode tube, the heater, H, heats the cathode, K, and expands the orbital radius of the electrons from its surface so they can be drawn to the plate, P.

If there are only the heater, cathode, and plate, the tube is called a diode as it has two active electrodes. (The heater is not considered when we number the active electrodes.) A diode, Fig. 6, will act as a check valve when placed in an AC circuit, allowing electron flow only during the half cycle when the plate is positive with respect to the cathode. The output is pulsating DC with zero voltage each time the plate is negative.

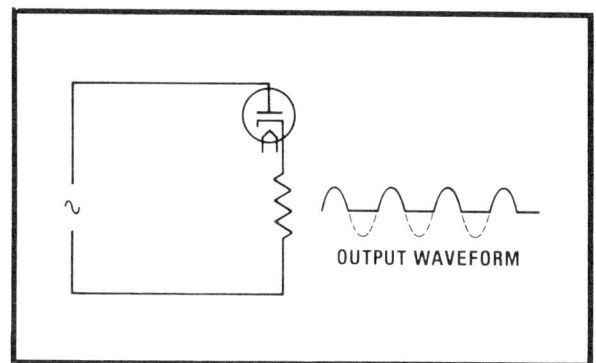

Fig. 6 A diode acts as a check valve in an AC circuit, allowing electron flow only when the plate is more positive than the cathode.

2. Triodes

A three-element tube, or triode, serves as an electron control valve and has an extremely fine wire spiraled between the cathode and the grid close to the cathode, Fig. 7. A small voltage placed on the grid has more influence on the electrons being emitted from the cathode than the higher voltage on the plate, farther away. In Fig. 8, the plate circuit consists of the plate, load resistor R_L, and the power supply, shown here as a battery E_B and the cathode. (It is assumed that the heater is operating.) If these were the only

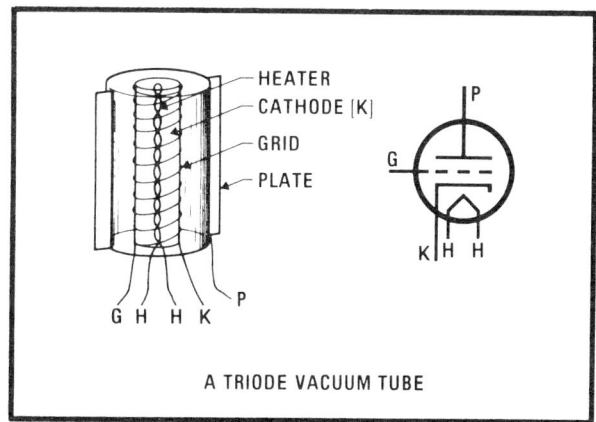

A TRIODE VACUUM TUBE

Fig. 7 A triode acts as a control valve, with a small potential on the grid controlling the flow of electrons between the cathode and the plate.

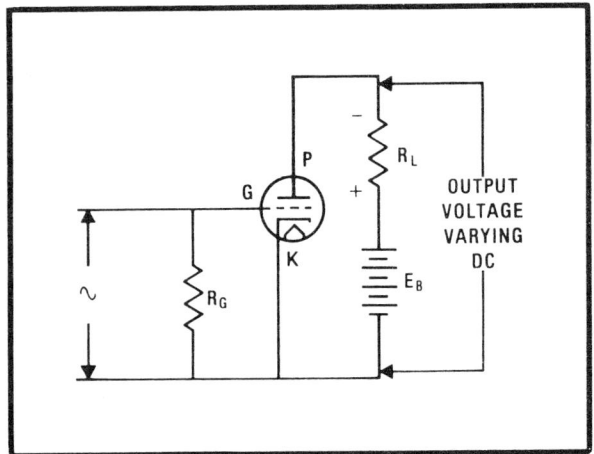

Fig. 8 An AC voltage applied to the grid resistor, R_G, will produce a varying DC output.

components in the circuit, electrons would flow from the negative terminal of the battery, across the space from cathode to the plate, back through the load resistor, to the positive terminal of the battery.

The input, or grid circuit, consists of a high-resistance resistor, R_G, between the grid and cathode, and the input signal across this resistor (R_G) applies a changing voltage to the grid. The high resistance of R_G prevents the tube taking more than an absolute minimum of current from the input circuit and gives the tube its high impedance characteristic. When the voltage on the grid is positive, the grid will attract electrons from the cathode, but since it is so extremely small, and the voltage on it is relatively low, most of the electrons will miss it and will be drawn to the plate. When the grid has a negative voltage on it, it will repel the electrons enough that they will stop entirely, and the tube is then said to be cut off.

3. Tetrodes

At high frequencies, the inter-electrode capacitance in a triode becomes so low that the output will feed back from the highly positive plate into the almost negative grid. This causes the tube to oscillate, or generate an unwanted AC voltage. To prevent this oscillation, an additional grid, called a screen grid, is built into the tube to neutralize the inter-electrode capacitive effect.

The screen grid is held at a high positive DC voltage, just a little lower than that of the plate,

but the capacitor C_{SG} places the screen grid at ground potential as far as AC is concerned. Now the inter-electrode capacitances in the tube are between the plate and the screen grid, and an AC fed back from the plate into the grid goes directly to ground and causes no problem. There is also a capacitive effect between the screen grid and the control grid, but since there is no AC on the screen grid, there can be no feedback into the control grid. The use of this additional grid makes a vacuum tube usable for some of the higher frequencies.

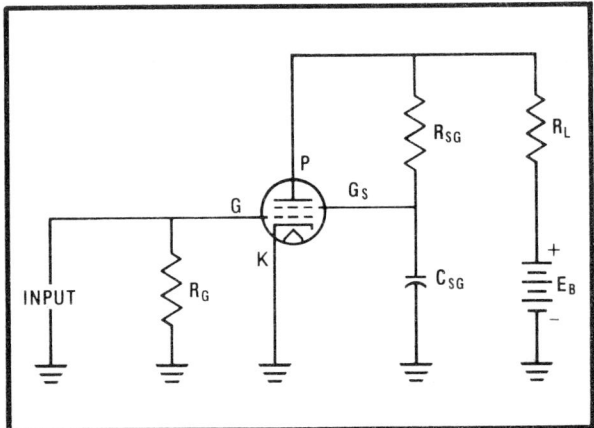

Fig. 9 A screen grid, G_S, placed between the control grid and the plate, minimizes the interelectrode capacitance and makes the tetrode usable for high frequencies.

4. Pentodes

At high operating power, the electrons that pass from the cathode through the control grid and screen grid of a tetrode tube may reach such high velocity that when they strike the plate some of them will bounce off and be attracted to the positively charged screen grid. This can cause excessive screen current and overheating of the screen grid. To prevent this secondary emission, as it is called, a third grid is installed in the tube, this one between the plate and the screen grid.

This grid is connected directly to the cathode in some tubes, and in others it is connected to ground outside of the tube. When electrons strike the plate and bounce off, instead of being picked up by the screen grid, they are suppressed by the negative field of the suppressor grid, back into the plate.

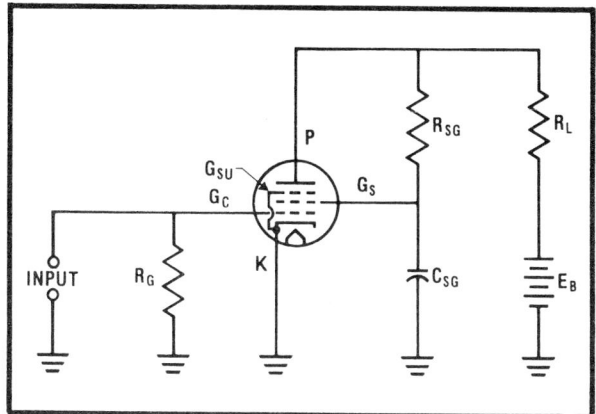

Fig. 10 The suppressor grid, G_{SU}, in a pentode vacuum tube suppresses secondary emission from the plate and minimizes screen current.

QUESTIONS:

5. Does a triode vacuum tube act as a high-impedance or a low-impedance device?

6. What is the purpose of a screen grid in a tetrode vacuum tube?

7. What is the purpose of a suppressor grid in a pentode vacuum tube?

B. SEMI-CONDUCTOR DIODES.

A conductor is a material whose atoms have their valence electrons, those in the outer orbit, so loosely bound to the nucleus that a relatively small force can cause them to move out and be replaced

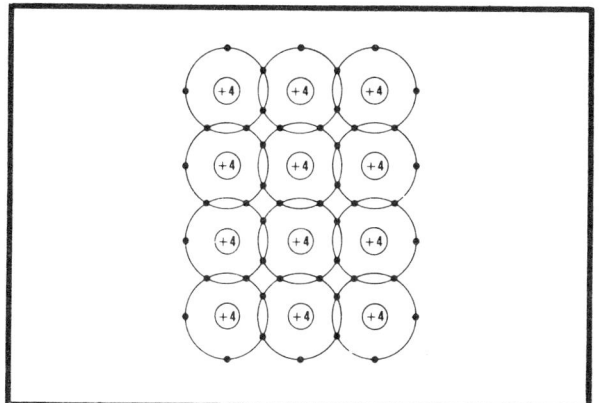

Fig. 11 All of the valence electrons in an insulator are tightly bound with covalent bonds, and the material will not accept electrons from an outside source.

by electrons dislodged from an adjoining atom. An insulator, on the other hand, has its valence electrons so tightly bound to those in adjacent atoms with covalent bonds that a high voltage is needed to dislodge and move them.

Silicon has four electrons in its outer shell and is a good example of an insulator. The valence electrons of one silicon atom are held tightly in covalent bonds with those of the adjoining atoms, and there are no loose electrons to give up, nor are there any vacancies to accept free electrons.

If the silicon is doped—having a few parts per million, of an element such as arsenic, bismuth or antimony, with five electrons in their outer orbit, mixed in,—there will then be extra, or free, electrons after all of the covalent bonds have formed. These electrons are free to move, and so the material is called a donor, or "N" type, material.

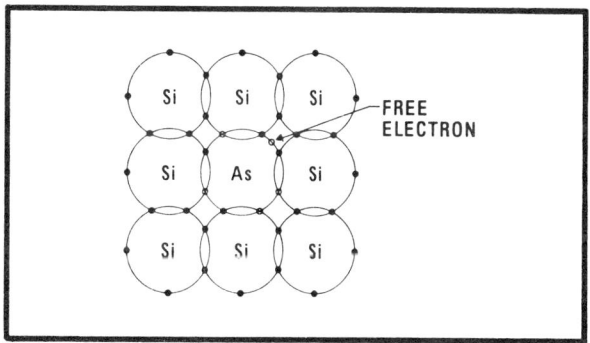

Fig. 12 An insulator, doped with an impurity having five valence electrons, produces an N material, a donor.

Elements such as boron, indium, or gallium have only three valence electrons, and when a few parts per million are alloyed with silicon, there will be areas where covalent bonds have not formed. These areas are called holes and will accept electrons from an outside source. Material doped in this way is called an acceptor, or "P" type, material.

N- and P-type silicon or germanium may be joined, either by a junction or a point contact, forming a semiconductor diode. To understand the principles, let's consider a junction formed by a piece of P and a piece of N silicon. The holes in the P material attract the electrons in the N material and, at the junction, they combine, or diffuse, leaving a depletion area where there are

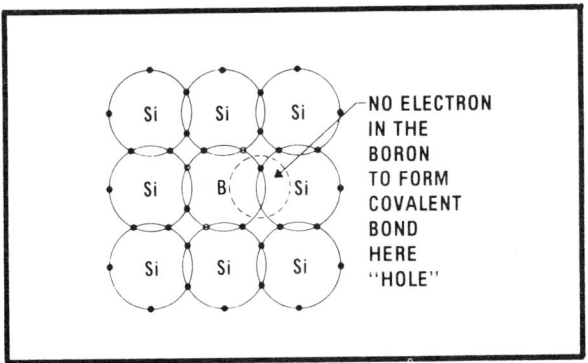

Fig. 13 If an impurity with three valence electrons is used to dope silicon, "holes" appear which will accept electrons. This is a P type material, an acceptor.

no more free electrons or holes. When the electrons have left the N material, its side of the junction has assumed a slight positive charge (lack of electrons); as the holes left the edge of the P material, its side assumed a negative charge. These charges constitute a barrier or potential hill, whose intensity is proportional to the width of the depletion area.

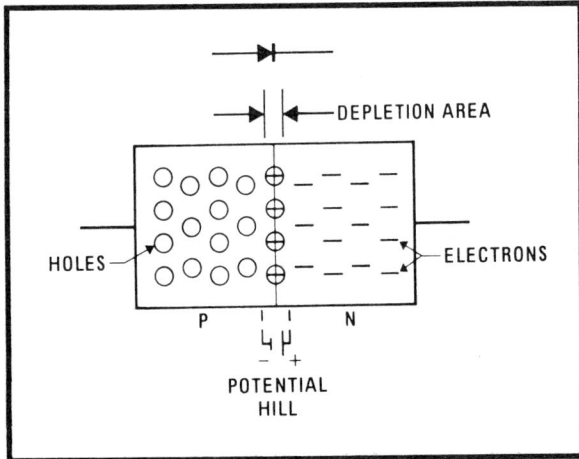

Fig. 14 When a piece of P and N silicon are joined, electrons and holes combine along the junction and form a depletion area, or a potential hill.

If a voltage source is attached to a semiconductor diode with its positive terminal to the P material and its negative terminal to the N, the diode is said to be forward biased, and electrons can flow through it. The negative potential of the battery has forced the electrons toward the junction, while the positive voltage forced the holes toward the junction. At the junction, the electrons and holes

combine, making room for more electrons to enter the N side. In the depletion area, holes and electrons are continually combining, and the area becomes extremely narrow. As a result, the barrier, or potential hill, is very small.

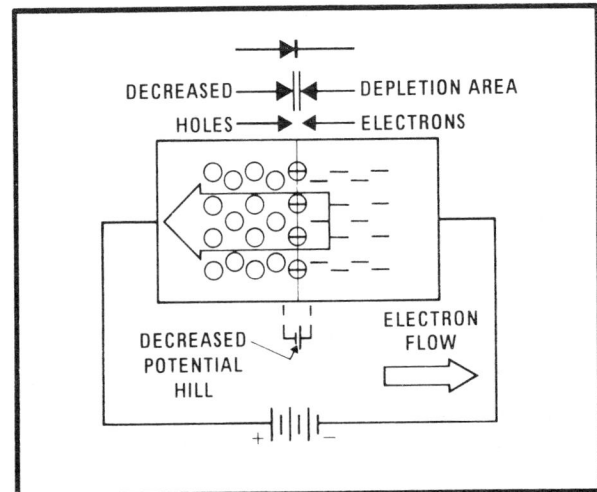

Fig. 15 A semiconductor is forward-biased when the positive terminal of the voltage source is attached to the P material and the negative to the N. Electrons can flow.

When the power source is turned around so that its positive terminal attaches to the N material and its negative terminal to the P, the electrons and holes will be attracted away from the junction; the depletion area will enlarge so much and the potential hill become so high that no electrons or holes will combine, and there will be no electron flow in the external circuit. The junction is said to be reverse biased.

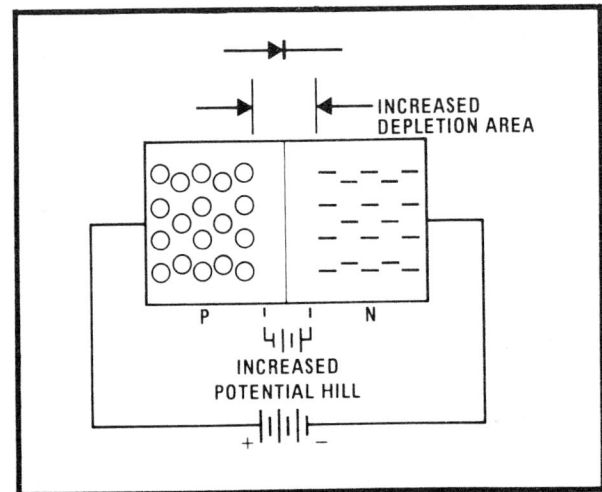

Fig. 16 When the semiconductor is reverse biased, no current can flow.

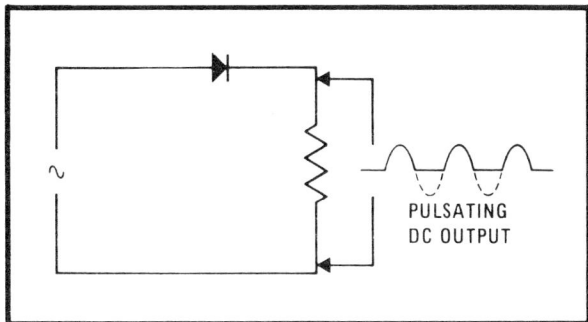

Fig. 17 A diode acts as a check valve when it is put in an AC circuit. It allows electron flow when forward baised but none when reverse biased.

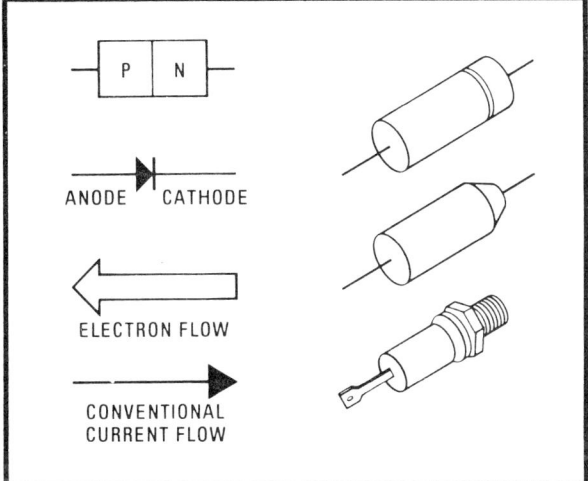

Fig. 18 Symbols and forms of semiconductor diodes.

One of the important applications for a semiconductor diode is its use as a rectifier in an AC circuit. In Fig. 17, a single diode is placed in series with a load resistor, across which we can measure the voltage. The diode acts as a check valve, allowing electron flow during the half cycle it is forward biased, and refusing flow during the half cycle when it is reverse biased.

C. ZENER DIODE

Fig. 19 is a curve representing the characteristics of a semiconductor diode, showing the way current flow varies as a result of the applied voltage.

When a diode is forward biased, its current flow is roughly proportional to the applied voltage; that is, the device is essentially linear. Above its

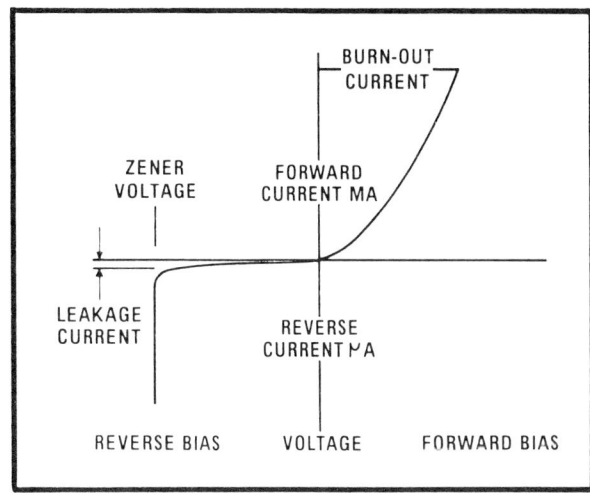

Fig. 19 Zener diode conduction characteristics.

burn-out point, however, so much current will pass that the diode will overheat and burn up. When the diode is reverse biased—that is, put into a circuit in such a way that the positive of the source is at the cathode—very little current will pass, usually in the realm of microamperes. This is called leakage current and is dependent, among other things, on the temperature of the diode.

Now, there is a rather sharply defined point of reverse voltage, called the zener voltage or the avalanche point, at which the diode stops acting as a check valve and breaks down, allowing current to pass in the reverse direction. You notice from the curve that when the zener voltage is reached, the current increases almost instantanously until it will burn out the diode if no form of limiting resistor is installed in the circuit.

Zener diodes are used in circuits for protection against overvoltage and, in a more familiar form to the A&P, as the sensing device in a transistor type voltage regulator.

QUESTIONS:

8. Will N-type silicon accept electrons, or does it have an excess of electrons?

9. What causes the ''holes'' in P-type silicon?

10. When a power source is attached to a semiconductor diode with its positive terminal to the N material, is the diode forward or reverse biased?

D. TRANSISTORS

Diodes serve well as check valves, but for the practical use of electron flow we must have control valves. So in much the same way that the triode vacuum tube opened up the field of electronics, so

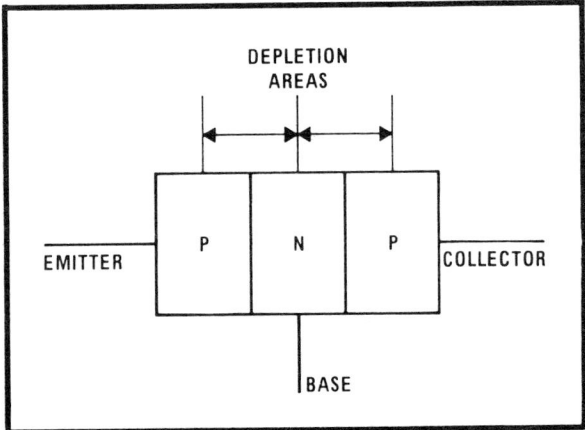

Fig. 20 Depletion areas exist across each junction in a PNP transistor.

has the triode semiconductor, the transistor, made possible the miniaturization of electronic devices.

A transistor is essentially either a sandwich of N silicon or germanium between two pieces of P material, or a piece of P between two pieces of N. The design and construction of these miniature

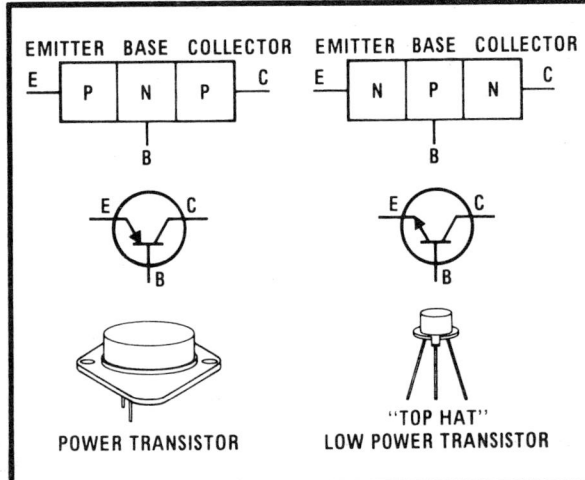

Fig. 21 Commonly used transistor symbols.

electron control valves is extremely complex, but as A&P technicians, we are concerned primarily with the practical aspects of their theory and with their practical uses.

For a transistor to conduct, the emitter-base junction must be forward biased and the collector-base, reverse biased.

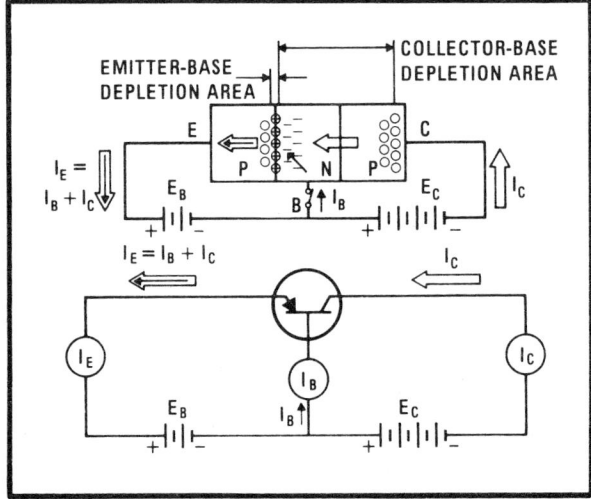

Fig. 22 The emitter-base junction is forward biased, and base current flows. By controlling the small base current, the much larger emitter-collector current is controlled.

A small amount of current flows between the emitter and the base, and this causes an extremely narrow depletion area. The relatively large reverse bias between the collector and the base forces electrons to the emitter-base junction, where they are attracted to the positive voltage source attached to the emitter. Since the base is so thin, the electrons which leave the collector, rather than returning to the positive terminal of the collector source, return to the positive terminal of the emitter source. A very small base current is all that is required to keep the

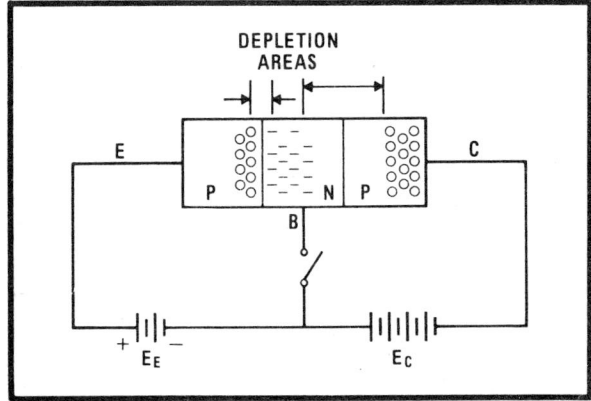

Fig. 23 When the base circuit is opened, no base current can flow and therefore no emitter-collector current.

emitter-base depletion area small enough that a much larger current can flow between the collector and the emitter.

When the base circuit is opened, Fig. 23, there is no longer a force to keep the emitter-base depletion area reduced; so there will be no attraction for the electrons from the negative terminal of the collector source across both depletion areas to the positive terminal of the emitter source. When no base current flows, collector current cannot flow.

When the base of a PNP transistor is made negative with respect to its emitter, base current will flow, and the transistor conducts between its collector and emitter. It is sometimes desirable, in a circuit, to have a positive signal voltage cause a transistor to conduct. For this reason, there are NPN transistors. NPN transistors are similar in almost every way to the PNP except for the arrangement of the doped areas. When biasing an NPN transistor, the voltage polarities are exactly opposite those of the PNP. For maximum conduction with both types, the emitter-base junction must be forward biased, and the collector-base, reverse biased.

QUESTIONS:

11. For conduction, is the emitter-base junction of a transistor forward or reverse biased?

12. What is the difference in the way a PNP and an NPN transistor are connected to their power supplies for best conduction?

E. SILICON CONTROLLED RECTIFIERS

Electrical circuits for lights and some types of motors may limit the current by dropping the voltage supplied to these devices. This requires a relatively large resistor and dissipates a lot of energy in the form of heat. It is possible to dim lights and control the speed of universal motors by using a silicon-controlled rectifier, SCR, rather than the resistor. An SCR decreases the amount of current supplied to the device, not by dropping the voltage and dissipating power, but by controlling the time in the cycle of AC the current is allowed to flow.

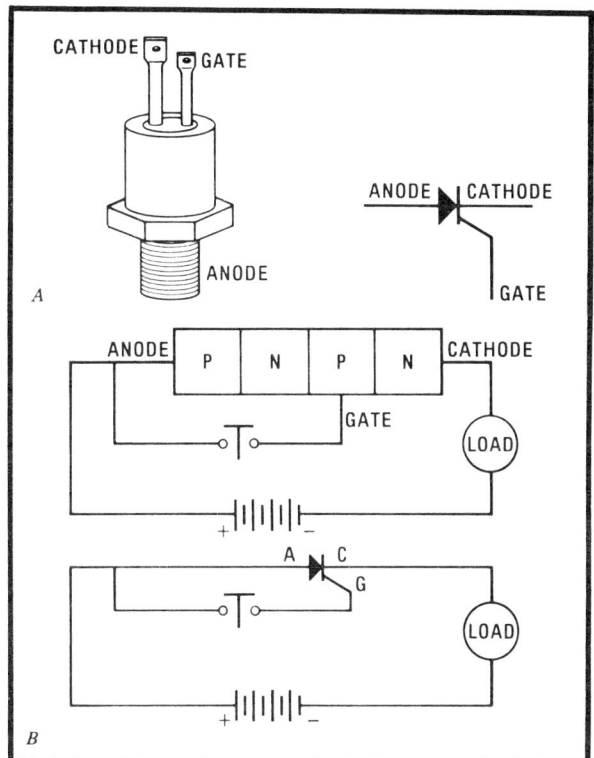

Fig. 24 Silicon Controlled Rectifier
A - Typical form and symbol of an SCR.
B - When a positive pulse is applied to the gate, both P-N junctions are forward biased and the SCR will conduct until the voltage is removed.

A silicon-controlled rectifier is similar to a silicon diode in its outward appearance, except for its extra terminal, the gate. Another difference is that the case of a stud type SCR is its anode, while on a regular diode, the case is the cathode.

An SCR is a special type of diode having three junctions, Fig. 24, two of these junctions are forward baised, and one reverse biased, so no electrons can flow through the load. If the gate is *momentarily* connected to the positive voltage at the anode, the reverse biased junction will become forward biased, and electrons will flow through the SCR. Once this flow has started, it will maintain the forward bias and flow will continue until the voltage across the SCR is removed. No further voltage need be applied to the gate.

The circuit of Fig. 25 shows the way an SCR can be used to produce controlled direct current from an alternating current source.

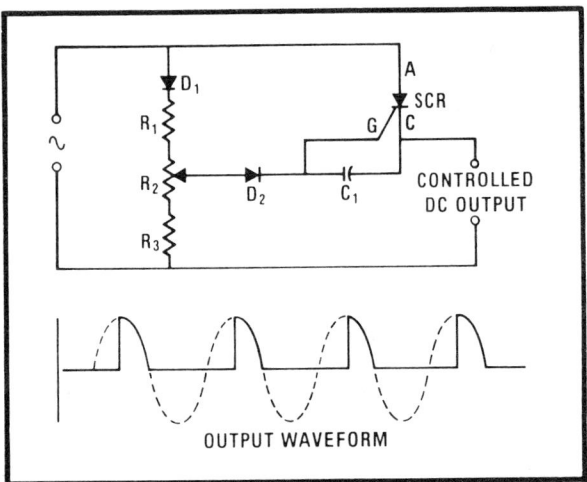

Fig. 25 *The position of control R_2 determines the time in the cycle the SCR begins to conduct. Current will flow only during the positive half-cycle after the SCR has been triggered.*

The diode D_1 and the voltage divider, R_1, R_2 and R_3, provide an adjustable direct current to trigger the SCR. Looking at the waveform controlled by the SCR, we see that there is no flow at all in one half of the cycle, that half when the anode is negative. Neither is there any flow during the half cycle when the anode is positive, *until* the voltage across the divider has risen sufficiently to charge the capacitor C_1 enough to trigger the gate of the SCR. When the voltage rises high enough, the gate is triggered, and the charged capacitor provides the needed pulse to start the SCR conducting. Once it starts conducting, it will continue until the supply voltage shuts off, or reverses, as it does in the next half cycle.

F. TRIACS

One of the limitations of a silicon-controlled rectifier is that it controls only one half of the cycle of AC. A triac, on the other hand, overcomes this by acting as though it were two SCRs connected side by side, in opposite directions. An SCR requires a positive pulse to trigger its gate, but a triac can be triggered by a pulse of either polarity, and the output waveform will appear as that in Fig. 26. For full power, the triac is triggered at the beginning of the cycle and all of the current flows; but if it is triggered later in the cycles as seen here, only about half of the current flows.

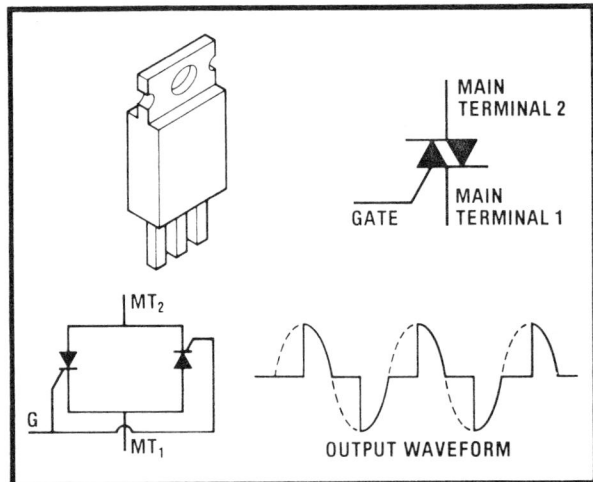

Fig. 26 *A triac may be triggered with a pulse of either polarity, and will conduct during the portion of the cycle after it has been triggered.*

QUESTIONS:

13. How can an SCR dim lights without dissipating a lot of power?

14. What kind of pulse, negative or positive, is needed to trigger an SCR into conduction?

15. What is an advantage of a triac over an SCR for dimming lights in an AC circuit?

G. FIELD EFFECT TRANSISTORS

A transistor is a *low impedance* device, which, by controlling the *current* flow into or out of the base, controls the flow of current between the emitter and the collector. A vacuum tube, on the other hand, is a *high impedance* device, and variations in *voltage* on the grid will control the flow of current between the cathode and the anode (the plate).

It is advantageous to have a solid state device that will control the emitter-collector current by the control of a voltage rather than current, and the Field Effect Transistor (FET) has been developed to do just this. In Fig. 27-A we see that an FET is constructed of a channel of either N- or P-type silicon, and sitting in this channel, like a valve, is the gate. One end of the channel is called the source and the other, the drain. An N-channel FET has a P-type gate, so when a positive voltage is applied to the gate, the FET is forward biased, and there will be a greater flow of electrons between the source and the drain. If a negative voltage is applied to the gate, the FET will be reverse biased, and the flow between source and drain will be pinched off.

The Field Effect Transistor multivibrator in Fig. 27-B produces a square wave output from direct current. Assume that when power is supplied to the circuit, FET_1 conducts more heavily than FET_2: the entire voltage drop now appears across R_4 which causes both ends of R_3 to be positive and removes any forward bias from FET_2. Now, with FET_2 not conducting, there is no voltage drop across R_4, and the gate end of R_2 will become negative, forward biasing FET_1 into full conduction. Now, depending on the time constants of R_2 and C_2, the gate of FET_1 will become positive and pinch off, forcing the drain end of R_1, and the gate of FET_2 to go negative. This turns FET_2 on to full conduction. The speed at which the FET's can turn on and off make the output of this circuit a square wave.

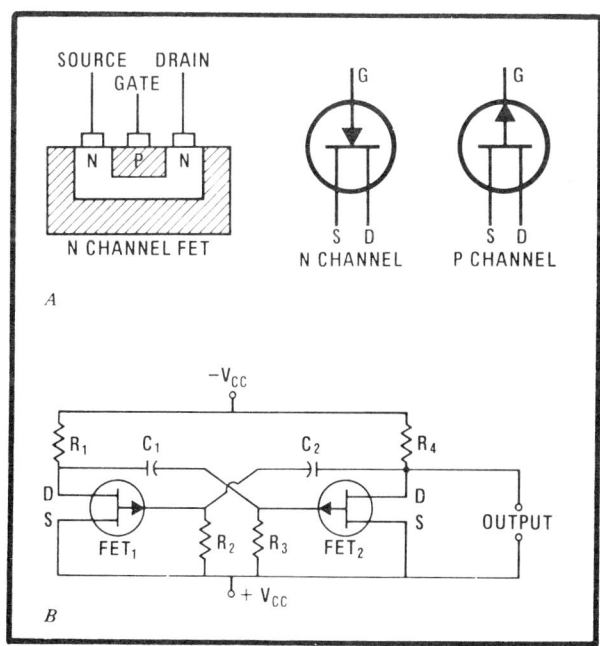

Fig. 27   Field Effect Transistor

A - A field-effect transistor will conduct between its source and drain when a forward-bias voltage is applied to its gate.

B - An FET multivibrator produces a square wave output.

H.   TUNNEL DIODES

As we saw earlier, if a diode is subjected to an increasing forward bias voltage, the current

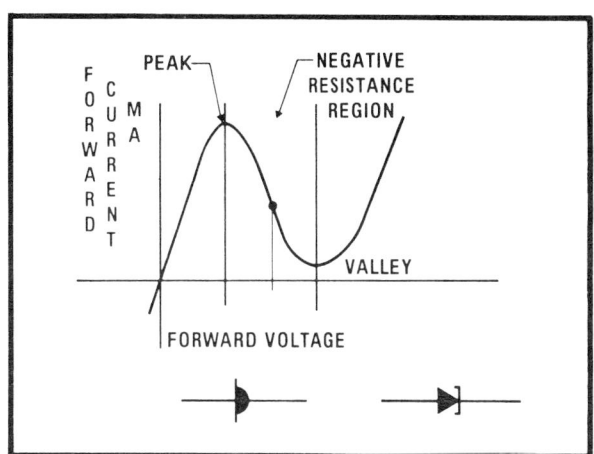

Fig. 28   A tunnel diode has an area of conduction in which an increase in voltage results in a decrease in current—negative resistance.

through it will increase, up to the point the diode burns out. This region of increasing current with increasing forward bias shows that the diode has a positive resistance, (increasing E, increases I). There are special diodes, however, called tunnel diodes, made up of heavily doped P and N sections, and these devices have the unusual characteristic of a region of negative resistance. Looking at Fig. 28, we see that as we increase the forward bias on the tunnel diode, the current will increase up to a point, the peak; and any increase in forward bias beyond the peak will put the diode into a region of negative resistance which causes the current to *decrease* with an *increase* in forward bias. (Increasing E, in this case, decreases I.) This condition continues to the valley, at which time the diode begins to act in a normal way.

Tunnel diodes may be used in oscillator circuits to supply the power lost, in order to sustain oscillation.

I.   UNIJUNCTION TRANSISTORS [UJT]

A unijunction transistor, sometimes called a double-base diode, is made of a single crystal of uniformly doped N-type silicon, and has contacts at each of its ends, and a small P-type emitter is located near its middle. A UJT acts like a good insulator until the voltage at the emitter becomes sufficiently high to trigger it, at which time it conducts with a minimum of resistance. UJT's are used in such circuits as relaxation oscillators where it is necessary to provide short, high-intensity current pulses when the control voltage rises to a given value. They may also be used to provide the gate pulses for silicon-controlled rectifiers or triac circuits.

In Fig. 29, when power is supplied to the circuit, C_1 begins to charge through R_1; and when the voltage across C_1 rises to the value required to trigger the UJT, Q_1 conducts and C_1 discharges through Q_1 and R_3. A pulse of energy is produced at the output sufficient to trigger an SCR or a triac or to supply a pulse for an integrated circuit counter. As soon as C_1 discharges, the UJT stops conducting and C_1 charges for the next pulse. The time constant of R_1 and C_1 determines the pulse rate of this circuit.

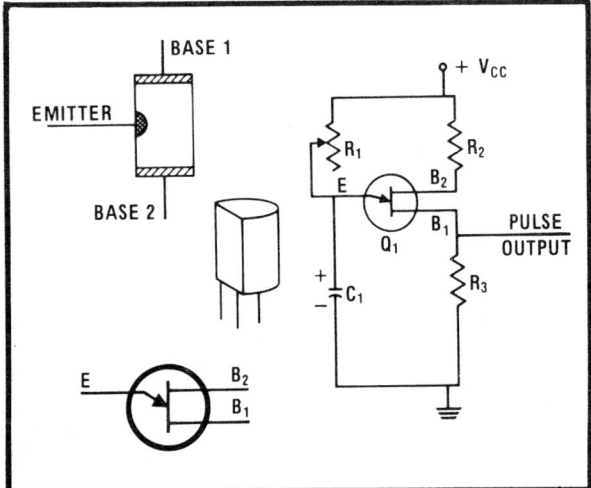

Fig. 29 A Unijunction Transistor produces a pulse output when the voltage between its emitter and base rises to a predetermined value.

QUESTIONS:

16. Is a Field Effect Transistor controlled by base current or by a voltage on its gate?

17. What is meant by negative resistance in a tunnel diode?

18. What causes a UJT to conduct?

J. OPTOELECTRONIC DEVICES

1. Photodiodes

As the science of electronics becomes more sophisticated, we are able to use the energy found in light for our devices and circuits. Light energy

is electromagnetic in nature, and one of its characteristics is that it will increase the reverse, or leakage, current in a semiconductor device. A photodiode is a special diode, so made that light shining through an aperture in its case will release free electrons into the depletion area and cause it to conduct.

In Fig. 30, the photodiode is installed in the coil circuit of a relay in such a way that current is normally blocked, but when sufficient light strikes the diode, current will flow and energize the relay, turning on or off any circuit desired.

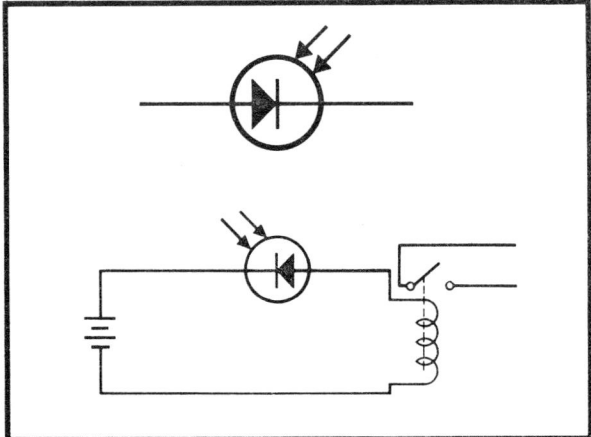

Fig. 30 A photodiode will conduct when sufficient light energy strikes its reverse-biased junction.

2. Phototransistors

A photodiode is limited in the amount of current it can pass, since it can not amplify; but if a

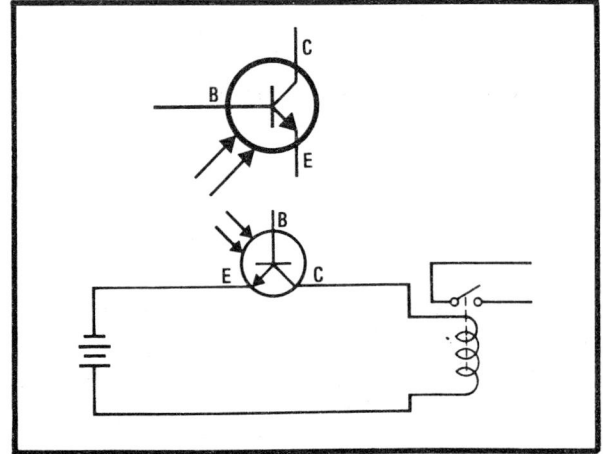

Fig. 31 A phototransistor will conduct when sufficient light strikes its emitter-base junction to provide a forward bias.

transistor could be activated by light, the circuit could carry much more current. In Fig. 31 a phototransistor is installed in a relay circuit, and biased in such a way that with no base current there will be no flow through the emitter and collector to the coil. But if light of sufficient energy strikes the collector-base junction, enough electrons will flow from the base to turn the transistor on, and the amplified current can flow from the emitter to the collector and energize the relay coil.

3. Light-emitting diodes

When a photon of light strikes a photo conducting material, an electron is freed; and, conversely, if a free electron in a piece of light-emitting semi-conductor material falls into a hole, a photon of light is generated. Light-emitting material has found almost immeasurable usage in the displays of digital electronic devices. When a light-emitting semiconductor is reverse biased, no current can flow and it remains dark; but when it is forward biased, it will emit light.

Light-emitting diodes find some of their most popular applications in the digital read-outs in instruments, calculators, clocks, and watches. In these devices, the LED's are usually arranged in seven bars, Fig. 32, and an Integrated Circuit causes the correct bars to light up to display whatever number is required.

QUESTIONS:

19. What causes a photodiode to conduct?

20. What is the difference between a photo diode and a phototransistor?

21. What is done to a light-emitting diode to cause it to conduct?

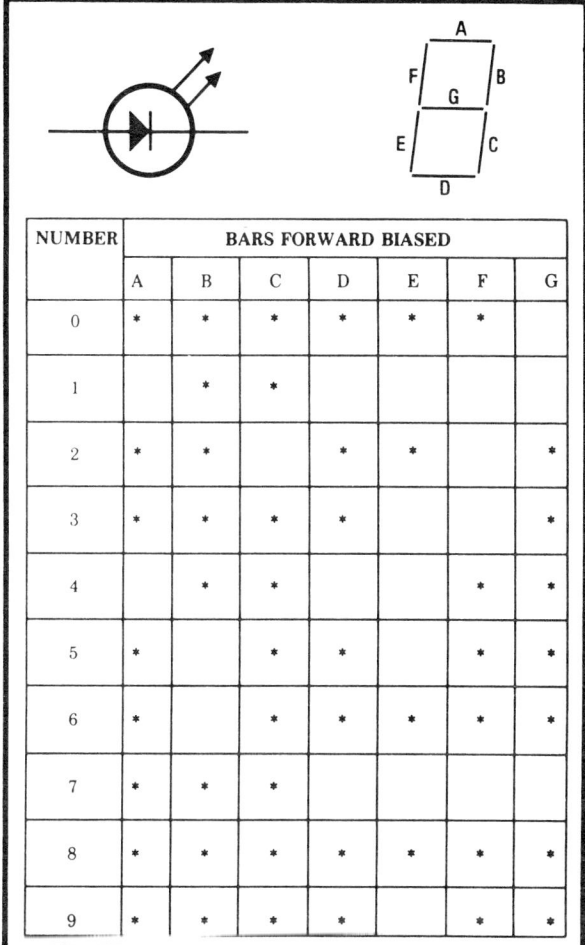

NUMBER	BARS FORWARD BIASED						
	A	B	C	D	E	F	G
0	*	*	*	*	*	*	
1		*	*				
2	*	*		*	*		*
3	*	*	*	*			*
4		*	*			*	*
5	*		*	*		*	*
6	*		*	*	*	*	*
7	*	*	*				
8	*	*	*	*	*	*	*
9	*	*	*	*		*	*

Fig. 32 Light emitting diodes emit light when current flows through them. Segmented LED displays are made up by forward biasing certain bars in the seven segment arrangement.

SECTION III:
Practical Circuits

A. RECTIFIER CIRCUITS

1. Half-wave rectifiers

One of the most commonly used circuits in aircraft electronics is the rectifier, a circuit using semiconductor diodes to change alternating current into direct current. Fig. 33 shows a circuit which will change AC into pulsating DC, but the disadvantages of this simple rectifier is its low efficiency—only one-half of each cycle is used, and there are very pronounced pulsations in its output waveform.

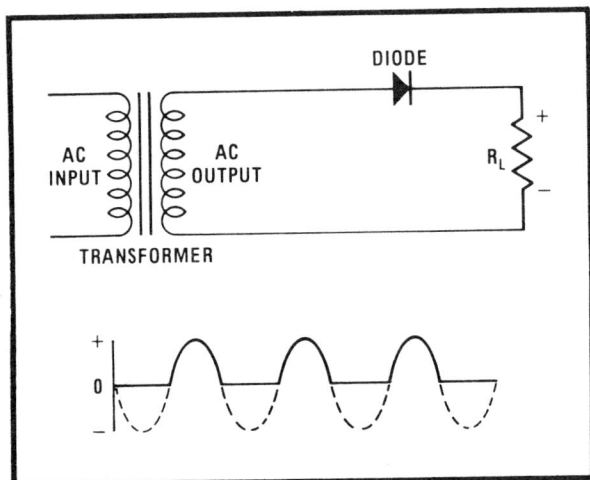

Fig. 33 *The half-wave rectifier used one check valve, [diode], in series with the load resistor R_L. The output voltage waveform is pulsating DC.*

2. Full-wave rectifiers

a. Two-diode

More efficiency can be obtained by using two diodes and tapping the secondary of the transformer. In essence this is two half-wave rectifiers with their output having the same polarity. During the half cycle when the top of the transformer secondary is positive, electrons will flow through the load resistor and diode 1 back to the top of the transformer. One-half cycle of the AC later, the bottom of the transformer secondary will be positive, and electrons will flow through

the load resistor and diode 2 to the bottom of the transformer secondary. This is still pulsating DC, but both halves of the wave are used and, since the frequency of the pulsations is twice that of the half wave rectifier, filtering is much easier.

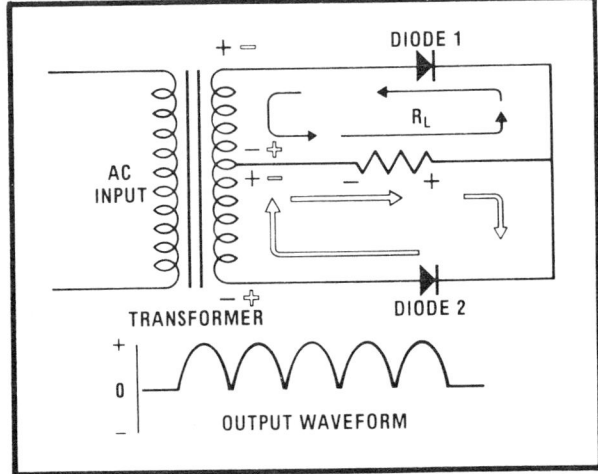

Fig. 34 *A full-wave, center-trapped rectifier produces pulsating DC with twice the frequency of the input AC.*

b. Four-diode-bridge-type

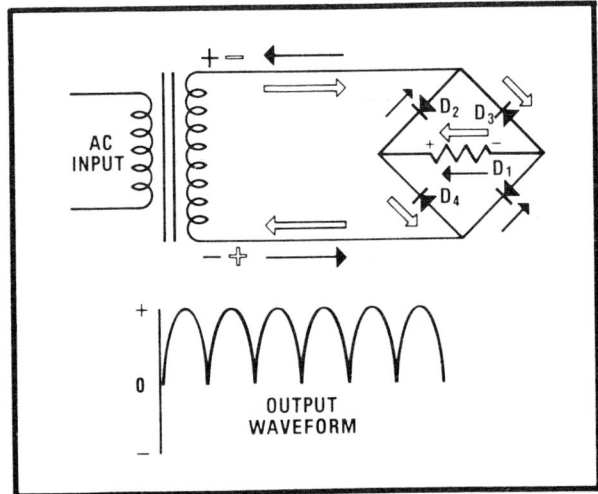

Fig. 35 *A four-diode, full-wave rectifier produces twice the output voltage of a two-diode rectifier which uses a center-tapped transformer.*

A rectifier circuit using four diodes makes it possible to use the entire transformer secondary. Fig. 35 is the circuit most commonly used for practical rectification of a single-phase AC.

When the bottom of the transformer secondary is negative, electrons leave it, going through diode 1, the load resistor, and diode 2 to the top of the transformer. During the next half-cycle, the electron flow is from the top of the transformer, through diode 3, the load resistor in the same direction as before, and diode 4, back to the bottom of the transformer. The output waveform is pulsating DC, but with a frequency twice the input AC frequency; the output amplitude is that put out by the complete secondary instead of half, as was the case with the two diode-rectifier.

c. Three-phase

Aircraft DC generators get a fairly smooth output by using a commutator and a brush-type mechanical rectifier, but alternators use solid state rectifiers to convert their three-phase AC into DC.

Fig. 36 shows a typical aircraft alternator and its six-diode rectifier.

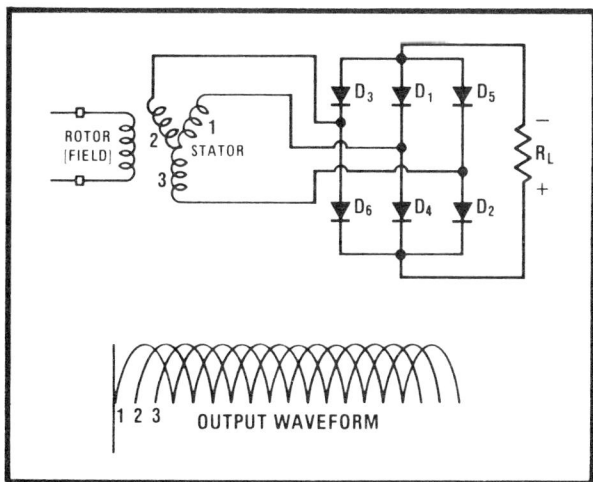

Fig. 36   Three-phase AC is rectified with a full wave rectifier using six diodes.

During the 120 degrees when phase 1 is negative, electrons flow from it through diode 1, the load resistor, and diode 2, to the phase 3 terminal. During the next 120 degrees of rotation, phase 2 will be negative and phase 1, positive; so electrons flow from phase 2 through diode 3, the load

resistor, and diode 4. The next 120 degrees, phase 3 will be negative and 2, positive; so the electron flow will be through diode 5, the load resistor, and back through diode 6. The output wave form shows the waves from each phase, but the resultant DC is actually just the top of the curves, since it never drops to zero as it does with a single-phase rectifier.

QUESTIONS:

22. What is one of the disadvantages of a half-wave rectifier?

23. What is the advantage of a full-wave rectifier over a half-wave?

24. What is the advantage of a four-diode full-wave rectifier over a two-diode full-wave recifier?

25. How many diodes are required to produce full-wave rectified three-phase AC?

B. AMPLIFIER CIRCUITS

An amplifier is a circuit that changes the amplitude of a signal; some amplifiers make the signal larger, while others make it less. We will look at three forms of transistor amplifiers and consider the way they change the output of the signal supplied to them, not only with regard to amplitude but also to phase.

In the previous section we studied the operation of a transistor in a circuit using two batteries. This is an awkward arrangement, so we will replace one of the batteries with a resistor and end up with the same results.

For a transistor to conduct, the emitter-base must be forward biased, and the collector-base reverse biased. The simple schematic of Fig. 37 shows a way of developing the proper bias voltages using only one power source.

1. Common-emitter amplifier

In addition to increasing the amplitude of the signal, a common-emitter amplifier inverts the phase. When an input signal is applied between the base resistor, R_B and ground, Fig. 38, it changes the bias on the emitter-based junction.

If the signal is negative, it increases the forward

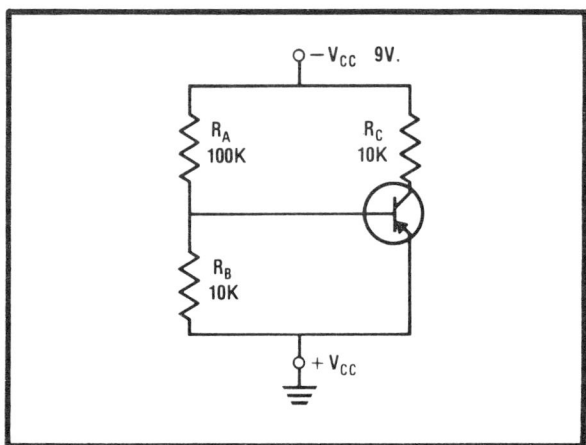

Fig. 37 By properly choosing the values of the resistors, a transistor may be biased for conduction with only one power source.

bias and increases the collector-emitter current through the transistor. When it is positive, it opposes the voltage drop across R_B and brings the base voltage closer to that of the emitter so that the transistor conducts less. (Of course, these values will be reversed for an N-P-N transistor.) When the collector current increases, the voltage drop across the collector resistor R_C increases, and since this voltage is opposite in polarity to that of the battery, the output will become less negative, or positive. A negative-going input voltage produces a positive-going output voltage.

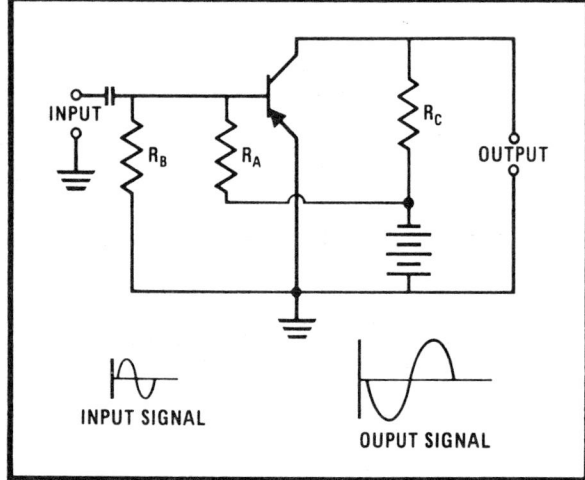

Fig. 38 The common emitter amplifier inverts the phase of the signal being amplified.

2. Common base amplifier

If a transistor is connected into an amplifier circuit

with the base common to both the input and output, a condition similar to that shown in **Fig. 39** results. The battery E_{EB} provides a forward bias for the emitter-base junction and a battery E_{CB} reverse-biases the collector-base junction. Resistor R_E is in series with the E_{EB}, and it is across this combination that the input signal is applied.

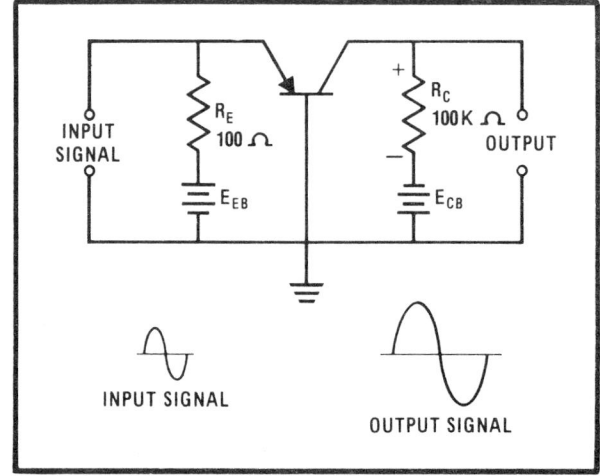

Fig. 39 The common-base amplifier has a low-impedance input and a high-impedance output, and does not invert the phase of the amplified signal.

Consider the portion of an AC input signal when the voltage is going positive; the forward bias will be increased, and the transistor will conduct more electrons in its collector-emitter circuit. This increase in collector current increases the voltage drop across R_C which, because of its polarity, will cause the output voltage to become less negative—or, in essence, positive. It is clear that this form of amplifier does not invert the phase of the signal as the common emitter amplifier did. The current through the emitter is the sum of the current that flows through the collector and the base, so there can be no current gain in a common base amplifier; in fact, the current amplification will always be less than one. The voltage gain, however, is another matter. The impedance of the output circuit is high and that of the input circuit is low so the current that flows through R_C will generate a much larger voltage drop than the voltage change across R_E.

3. Common collector amplifier

It is sometimes desirable to use a transistor circuit to transform a high impedance input into a low impedance output, and this may be done by using a common-collector amplifier circuit shown in Fig. 40.

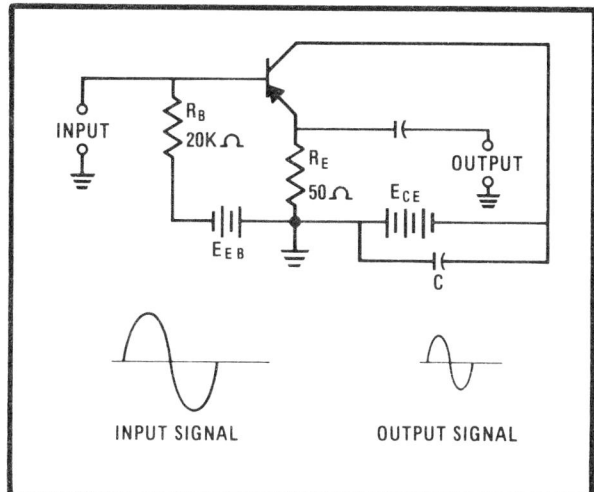

Fig. 40 *The common collector amplifier has a high-input impedance, a low output impedance, and does not invert the phase of the signal being amplified.*

In this circuit, battery E_{EB} provides the forward bias for the emitter-base junction, and the input signal is applied across this battery and resistor R_B. A negative-going signal will increase the forward bias, increasing the collector-emitter current and making the top of the resistor R_E more negative with respect to ground; and so the output voltage follows the input. Base resistor R_B has a high resistance and therefore only a small base current is required to control a rather large emitter current, and the low resistance of emitter resistor R_E provides a rather small voltage change. To summarize, the common collector amplifier has a voltage amplification of less than one, a rather high current amplification, and it

TYPE OF AMPLIFIER	IMPEDANCE	VOLTAGE GAIN	CURRENT GAIN	POWER GAIN	PHASE
COMMON EMITTER	INPUT: FAIRLY HIGH OUPUT: FAIRLY HIGH	RELATIVELY LARGE	RELATIVELY LARGE	LARGE	INVERTS PHASE
COMMON COLLECTOR	INPUT: HIGH OUTPUT: LOW	ALWAYS LESS THAN ONE	RELATIVELY LARGE	RELATIVELY LARGE	OUTPUT SAME AS INPUT
COMMON BASE	INPUT: LOW OUTPUT: HIGH	LARGE	ALWAYS LESS THAN ONE	RELATIVELY LARGE	OUTPUT SAME AS INPUT

Fig. 41 *Characteristics of transistor amplifiers*

does not invert the phase of the amplified signal. It provides an impedance step-down.

Fig. 41 summarizes the three most common types of transistor amplifiers, showing their voltage and current gain characteristics, their impedance relationships, and the effect of amplification on the phase of the signal.

QUESTIONS:

26. Does a common emitter amplifier invert the phase of the signal being amplifed?

27. Would a common collector amplifier be used in a circuit requiring a high or a low impedance input?

28. Does a common base amplifer invert the amplified signal?

C. OSCILLATOR CIRCUITS

Rectifiers convert alternating current into direct current; amplifiers change the value of either current, voltage, impedance or phase; and oscillator circuits make alternating current from DC. Rotary generators normally produce a nice smooth sine wave AC, but oscillators can produce almost any wave form required by electronic circuits. Before discussing oscillators, it is wise to consider some of the more common waveforms that may be encountered.

1. Wave forms

a. Sine wave

The waveform we normally think of when we consider AC is that generated by a conductor rotating in the uniform magnetic field. The voltage rises smoothly to a peak and drops to zero; then to a peak in the opposite direction, and again back to zero.

The value of voltage at any instant is equal to the peak voltage times the sine of the angle through which the conductor has turned. For example, in Fig. 42A, at 30° the voltage will be .50 times peak value, at 60° it is .866 times peak, and at 90° the voltage will be at its peak value.

b. Square wave

Oscillators that alternately allow a flow of electrons and then stop the flow, produce a wave form similar to that of Fig. 42B. In this instance the output is pulsating DC, but if this were passed through a transformer or coupled with a capacitor, it would produce AC, one polarity as the voltage rises, and the opposite polarity as it drops.

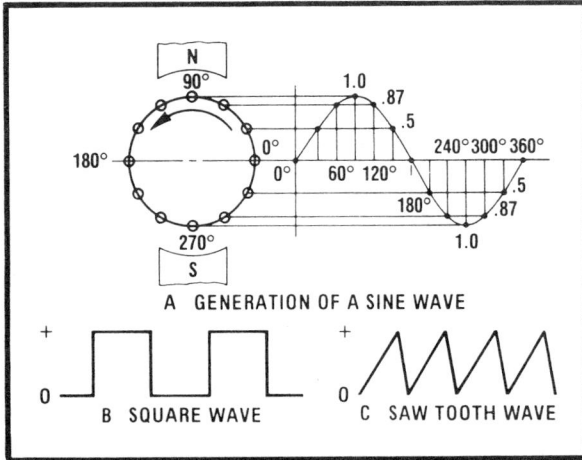

Fig. 42 Typical electrical waveforms.

c. Sawtooth wave

Relaxation oscillators produce a waveform similar to Fig. 42C. The voltage rise is relatively slow from zero to the peak at which time it drops to zero, rapidly. This form of wave is generated when a capacitor charges through a resistor to the ignition voltage of a neon bulb across the capacitor, Fig. 43.

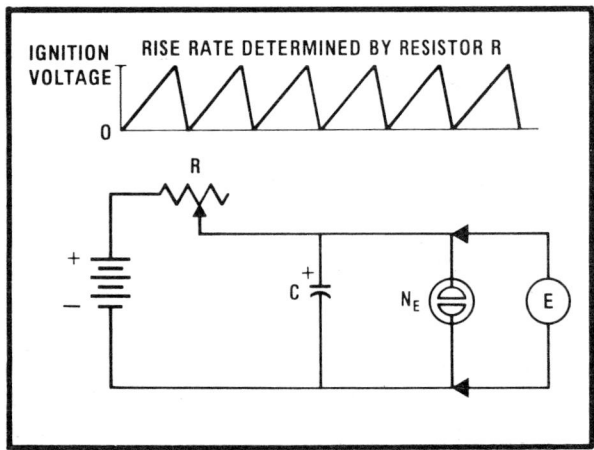

Fig. 43 Relaxation oscillator and its sawtooth waveform.

When the ignition voltage is reached, the bulb conducts, short circuiting the charged capacitor, and the voltage drops rapidly to almost zero. The bulb goes out, and the voltage again rises.

2. Electronic oscillation

a. Hartley oscillation

[1] Series

For electronic oscillation to occur, two conditions must be met; there must be amplification, and there must be feedback from the output circuit into the input.

Before we can fully understand oscillation, we must review one of the basic principles of AC electricity, that of resonance. If we have a coil and a capacitor connected in parallel, we have what is known as a tank circuit. In Fig. 44A, if we place the switch in position A, the capacitor with charge to the voltage of the battery, and no further current will flow. Then if the switch is placed in the open position, B, the capacitor will remain charged, with the electrical energy stored in the form of electrostatic fields within the capacitor.

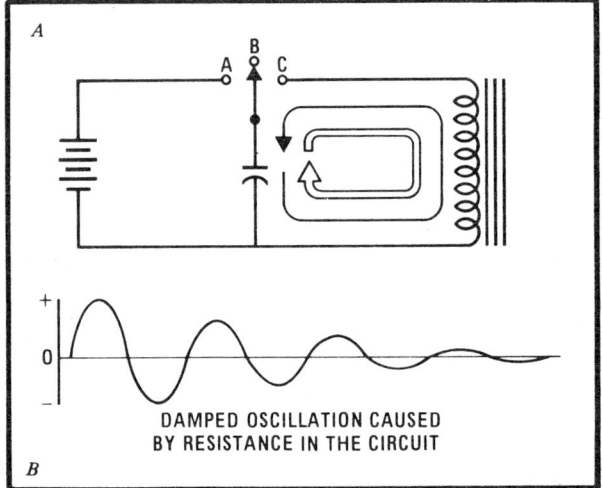

Fig. 44 Unless energy is added to an oscillator circuit to replace that lost in the resistance, the oscillations will die out.

When the switch is placed in position C, the capacitor will discharge throught the coil, and the energy in the electrostatic field will be transformed and stored in the form of an electromagnetic field around the windings of the coil. When the capacitor is completely discharged, the current will stop flowing, and the electromagnetic field will collapse, pushing electrons back into the capacitor on the plates opposite those originally holding them. When all of the energy has dissipated from the coil, the capacitor is charged, and it immediately begins to discharge through the coil again. If there were absolutely no resistance in the circuit, and the values of capacitor and coil were properly chosen, the charging and discharging or oscillation would continue indefinitely. But, unfortunately, there is always some resistance, and it damps the oscillations, eventually causing them to stop altogether.

In order for oscillation not to stop, the exact amount of energy which is lost must be restored. This is done by a circuit such as the Hartley oscillator of Fig. 45.

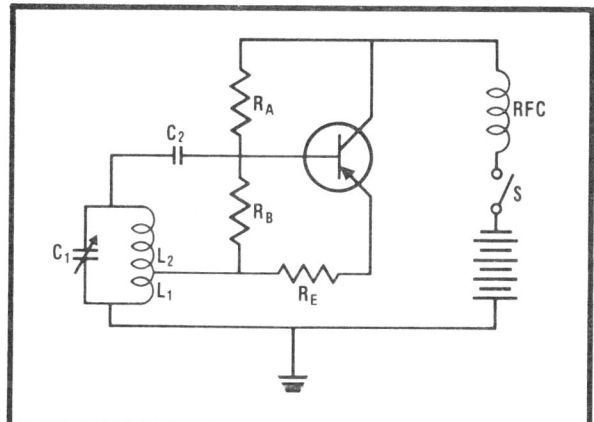

Fig. 45   Series fed Hartley oscillator.

When switch S is closed, electrons begin to flow in the circuit, biasing the transistor through the voltage divider, R_B and R_A, and flowing through L_1, the lower half of the center-tapped coil. As the current *rises* in this coil, it induces a voltage into the upper half of the coil that charges capacitor C_2 in such a direction that it increases the forward bias on the transistor, further increasing the electron flow. By the time the circuit is saturated—that is, there can be no further *increase* in electron flow—there is no more voltage induced in L_2 from L_1, and all of the energy stored in the tank circuit is the electrostatic fields in the capacitor C_1. Since there is no more force to push electrons into the capacitor it will discharge, and the energy lost will be made up by that stored in capacitor C_2 as it discharges. As C_2 is discharging, the forward bias decreases, and the transistor begins to conduct less, causing a decrease in the current in L_1. This change induces

a voltage into L_2 which causes it to charge C_2 in such a direction that it decreases the forward bias, and eventually reverse-biases the transistor, driving it all the way to cut-off. When all the flow is stopped, C_1 and C_2 will both be fully charged and the procedure will repeat itself. The coil, RFC, is a radio frequency choke which allows the flow of DC or audio frequency AC, but stops the flow of radio frequency AC. The frequency of the oscillations is determined by the inductance and capacitance in the tank circuit. Increasing the product of L and C will decrease the frequency, and decreasing it will increase the frequency of the AC produced. The frequency in hertz may be found by the formula:

$$FR = \frac{1}{2\,\pi\sqrt{LC}}$$

Here 2π is a constant; 6.28, L is the inductance of the coil in henries; and C is the capacitance of the capacitor in farads.

[2] Shunt

Another form of Hartley oscillator is the shunt-fed circuit in which the DC and the AC have separate but parallel paths. When the switch S, Fig. 46, is closed, electrons begin to flow through R_C and the collector-emitter to ground.

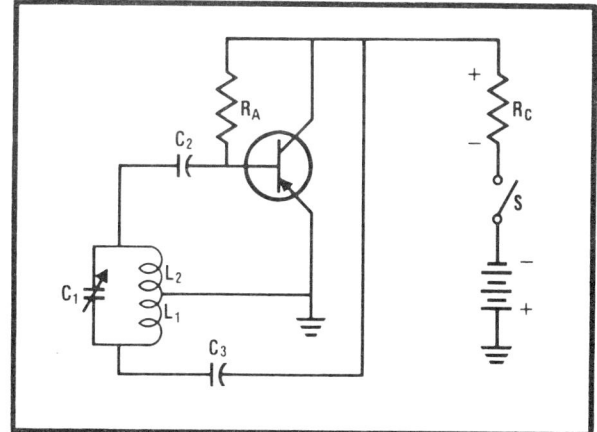

Fig. 46   Shunt fed Hartley oscillator.

Base current also flows through R_A and the emitter to ground, and this provides the needed forward bias for conduction. Now, as the current rises, the change is transmitted through C_2 to the base where it increases the forward bias until the transistor passes all of the current it can and it is saturated. At saturation the current can change no more, and C_1 and C_2 are fully charged. They will

3-6

now begin to discharge, with C_2 supplying the energy lost in the resistance of the circuit. As C_2 discharges, the transistor has its forward bias decreased so it begins to conduct less. This change passes through C_3, and the voltage it induces into L_2 drives the transistor to cut-off.

[3] Crystal

The product of the inductance and capacitance determine the frequency of oscillation, but if a more accurate control is required, a crystal may be inserted into the feedback circuit as shown in Fig. 47. Now, we must remember that all mechanical objects have a natural vibratory frequency; this means that a given mass and configuration of a material will vibrate at only one frequency. A small piece of a crystalline material such as quartz is held between two plates in such a manner that it is free to flex. When a pulse of voltage is applied between the plates, the crystal will vibrate at its natural frequency; as it vibrates, because of a characteristic know as the piezoelectric effect, it will generate a voltage each time it flexes.

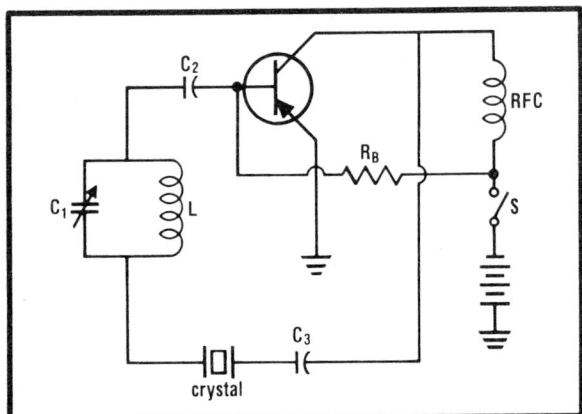

Fig. 47   Crystal controlled oscillator.

This crystal is put into the feedback portion of the circuit, and every time the voltage changes, it excites the crystal so it will send an accurately timed pulse into the circuit, sustaining the oscillation at a specific frequency. The crystal acts as though it is a series resonant circuit, passing only the frequency for which it is cut. The LC tank circuit must be tuned to the frequency of the crystal for sustained oscillation.

QUESTIONS:

29.   What causes the immediate drop in the

voltage across the capacitor in a simple relaxation oscillator?

30.   What two requirements must be met for sustained oscillation by an electronic oscillator?

31.   What determines the frequency of oscillation of a Hartley oscillator that does not use a crystal?

b. Multivibrator oscillator

Multivibrator oscillators are one type of electronic circuit which is becoming more popular as their applications in computers increase. In a multibrator circuit there are two transistors which alternately conduct. One type, the astable, or free-running, oscillator produces a square wave output. One transistor conducts; then, governed by the time constant of the capacitor and resistors, will abruptly and automatically shut-off. This causes the other transistor to conduct. In a bistable multivibrator, one transistor conducts until a pulse of energy causes it to turn off and allows the other to conduct. Either of them can conduct, but when one is conducting the other is driven to cut-off. In a monostable multivibrator, one transistor will conduct until a pulse of energy turns it off and the other one on. After a given time interval, again depending on the time constants of the circuit elements, it will swap back to the conduction of the first transistor. Multivibrators may use vacuum tubes instead of transistors, but in this book we are concerned primarily with the smaller, more simple solid state components.

[1] Astable multivibrator

An astable, or free running multivibrator, Fig. 48 is one in which the conduction of one transistor drives the other to cutoff, and when this condition stabilizes, depending on the time constants of the circuit, the two transistors swap conditions. This swapping out continues as long as power is supplied to the circuit.

In this circuit, when switch "S" is closed, Q_1 and Q_2 both begin to conduct, but let's consider that Q_1 conducts the most at the start: The collector end of R_1 will assume a positive charge because of the voltage drop across it, and this brings the base of Q_2 near enough its emitter voltage that it shuts

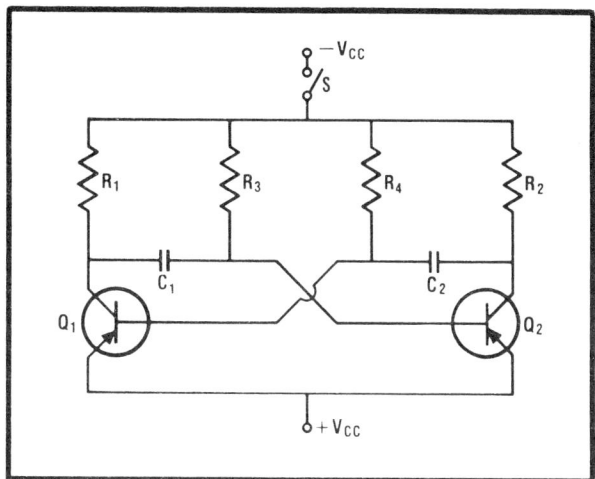

Fig. 48   Astable, or free running multivibrator.

off. With no flow through R_2, the base of Q_1 goes almost fully negative and Q_1 conducts as much as it can. (It goes to saturation.) With Q_1 conducting at saturation and Q_2 cut off, the reversal begins. Base current from Q_1 starts to charge C_2 which decreases the forward bias on Q_1. The flow through Q_1 decreases, and C_1 begins to discharge, allowing the base of Q_2 to become negative enough to start it conducting. Q_2 will go to saturation and Q_1 to cutoff, and the cycle will be ready repeat itself.

[2] *Bistable*

A bistable multivibrator finds many applications as a flip-flop circuit in modern digital computers. In Fig. 49, either transistor Q_1 or Q_2 will conduct, but not both at the same time. In order to switch from one to the other, a pulse of energy must be

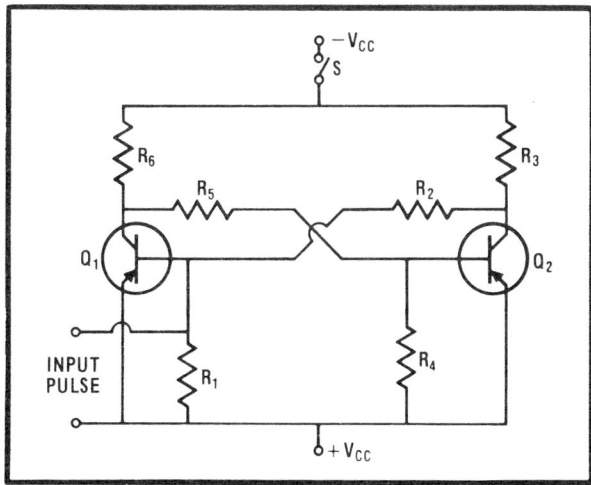

Fig. 49   Bistable multivibrator

applied to the circuit input. If the pulse is negative, Q_1 will conduct and if positive, Q_2 will conduct. The transistors will change conduction only when the pulse is applied to the input.

Let's assume that power is supplied to the circuit and a negative pulse is applied to the input across R_1. The base of Q_1 is driven negative with respect to its emitter, forward-biasing it, and Q_1 starts to conduct. The voltage drop across R_6 keeps the base of Q_2 positive enough to hold it shut off, and since there is not enough voltage drop across R_3 to reverse-bias Q_1, it will continue to conduct. Now, if a positive pulse is applied to the input, Q_1 will shut off, and since there is no longer a voltage drop across R_6, Q_2 will be forward-biased into conduction. The voltage drop across R_3 reverse biases Q_1 and holds it cut off until another pulse is applied to the input.

[3] *Monostable*

The monostable multivibrator is one in which Q_1 will conduct until a pulse of energy is supplied to the input, and then Q_1 will quit conducting and Q_2 will start. Q_2 will conduct for a time period determined by the time constants of the circuit, and then it will stop, and Q_1 will resume conduction until another pulse is applied to the input.

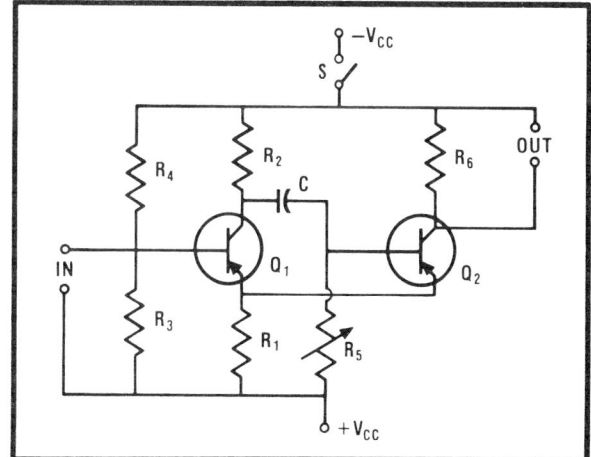

Fig. 50   Monostable multivibrator.

In Fig. 50, when the power is applied Q_1 will conduct because of the forward bias supplied by the voltage drop across the divider R_3 and R_4. There is no forward bias for Q_2, so it will not conduct.

Now when a positive pulse is applied to the input,

across R_3, Q_1 will be reverse biased and will stop conducting. The voltage divider R_5, C, and R_2 provides a forward bias for Q_2, and it conducts, and will continue conducting, until the capacitor C charges and removes the forward bias. With Q_2 no longer conducting, there will momentarily be no voltage drop across R_1, allowing Q_1 to regain its forward bias and again conduct until another positive pulse is applied to the input. You will notice that the output is across Q_2's collector resistor R6, so, when a positive pulse is applied to the base of Q_1, there will be a voltage drop across R_6 of a duration depending on the time constants of the circuit.

QUESTIONS:

32. What is meant by an astable multivibrator?

33. What is meant by a bistable multivibrator?

34. What is meant by a monostable multivibrator?

D. FILTER CIRCUITS

Rectifiers, either center-tapped full-wave or bridge type full-wave, produce pulsating direct current; and to be useful in circuits requiring pure DC, it must be filtered to smooth it out.

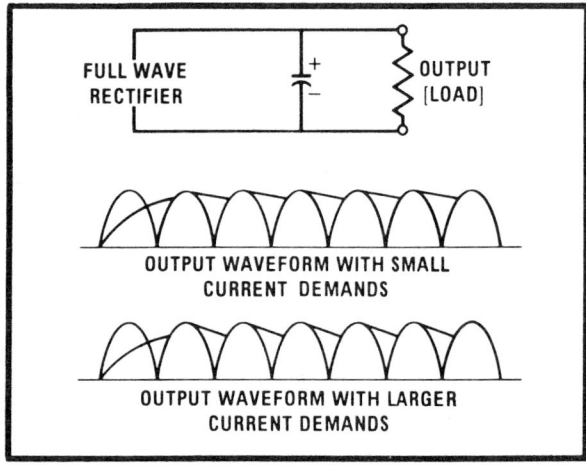

Fig. 51 The effectiveness of a capacitor input filter is determined by the current demands of the load.

A capacitor input filter of Fig. 51 partially smoothes the output. It does this by the capacitor charging up when the voltage rises and discharg-

ing when the voltage drops. As it discharges it supplies electrons to the load, minimizing the drop. When the load requires a small amount of current from the output, the waveform will be fairly smooth DC; but as the output demands are increased, the capacitor is no longer able to supply enough electrons to keep the voltage constant and a larger amount of ripple will show up in the output.

Instead of putting a capacitor across the load, if a choke coil is placed in series with the load, better filtering will be achieved, but there will be a somewhat lower output voltage. The inductance of the coil opposes the rise in voltage by generating a counter voltage, and as the voltage drops, the energy stored in the magnetic field is returned to the circuit.

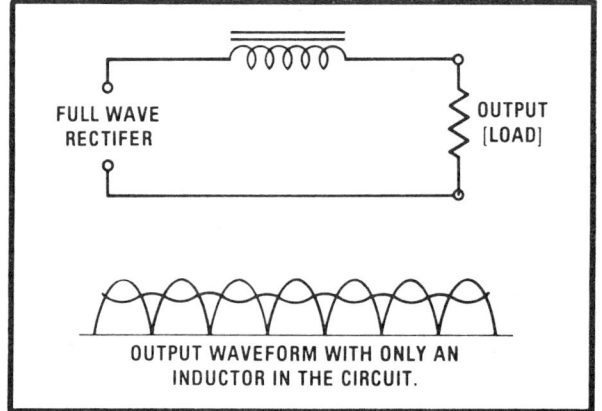

Fig. 52 A choke input filter tends to level out the peaks and valleys of the pulsating DC.

If a choke and a capacitor are both used in the filter circuit, with the choke in series with the load and the capacitor across the load, an L-filter is formed which has good regulation and little ripple.

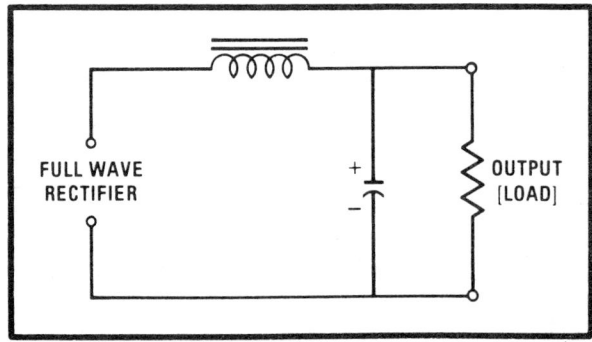

Fig. 53 An L type choke input filter has good regulation with little ripple.

An increase in output voltage may be obtained by using a capacitor input pi (π) filter, named after its similarity in shape to the Greek letter. The input capacitor C_1 has a low impedance to the ripple frequency of the rectifier output, and passes most of the ripple off to ground; while the L filter that follows smooths out the remainder of the ripple.

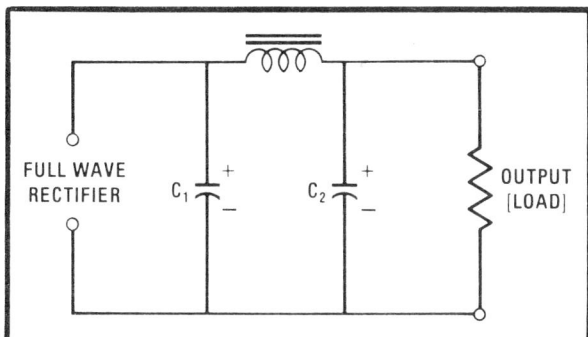

Fig. 54 *The capacitor-input pi filter is actually a capacitor, followed by an L filter in series. Pi filters are good primarily for low power applications.*

Better regulation of a power supply may be had by putting a bleeder resistor across the output of the filter.

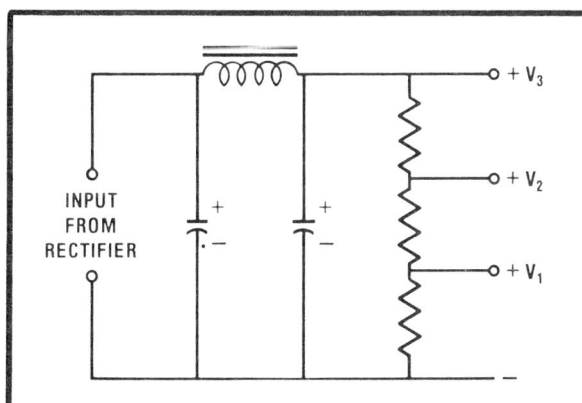

Fig. 55 *A bleeder resistor across the output of the filter improves the regulation, and discharges the capacitor when the power is turned off.*

The resistance of the bleeder should be such that it will pass about ten percent of the rated current of the supply, and in doing this, it provides the initial voltage drop, making the regulation for the load much better. The energy stored in the magnetic field of the choke coil because of the current improves its filtering action. But probably the most important function of the bleeder resistor

is to discharge the filter capacitors when the power supply is turned off. These capacitors usually have a large capacity and are of such good construction that they will maintain their charge for quite some time; and if it were not for the bleeder, they could be hazardous. Bleeder resistors may be tapped to serve as a voltage divider.

E. VOLTAGE MULTIPLIERS

1. Half-wave voltage doubler

It is often needful to increase the voltage in a circuit without using a transformer, and this can be easily done with a couple of diodes and two capacitors. In Fig. 56A the half cycle of input AC is seen when capacitor C_1 is charged to full line voltage with the polarity shown. No current can

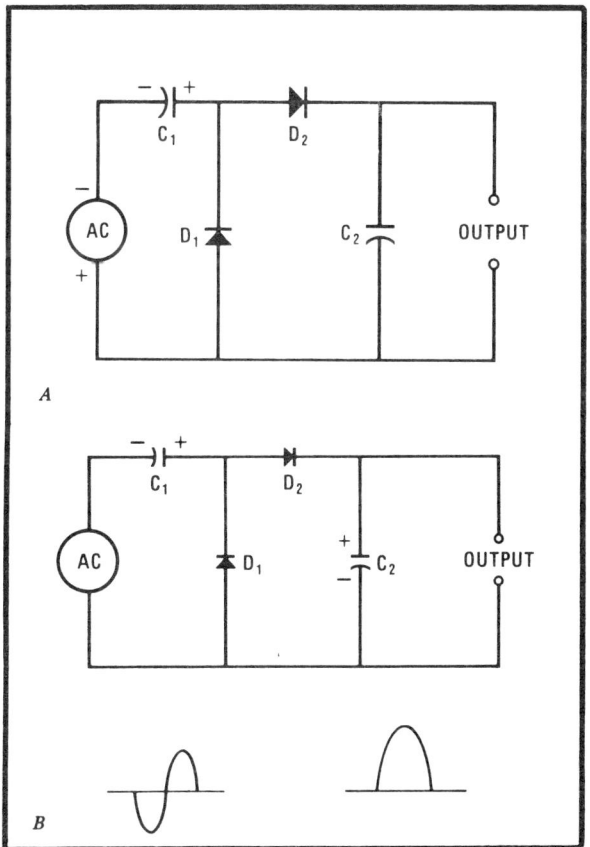

Fig. 56 *Half-wave voltage doubler*
A - *Here capacitor C_1 is charging to the full line voltage.*
B - *During this half-cycle, C_1 and the line voltage are in series across C_2, producing an output that is twice the line voltage.*

flow into C2 because of the blocking effect of D2. During the next half-cycle, Fig. 56B, C_1 and the input voltage have the same polarity and are in series across C_2. So the output will be half-wave rectified, but its voltage will be approximately twice that of the input.

2. Full-wave voltage doubler

The addition of one more capacitor can give an output that is twice the input. Rather than only one half of the cycle appearing, however, it will produce both halves. In Fig. 57A, during the half-cycle shown, capacitor C_1 charges to full line voltage through D_1; and during the next

half-cycle, C_2 charges to full line voltage through D_2 as is seen in 57B. C_1 and C_2 are in series, across C_3, so the output across C_3 is a full-wave rectification of the AC input with twice its voltage.

F. VOLTAGE REGULATORS

Many electronic circuits require that the output voltage not vary as the load changes, and in order to accomplish this, a zener diode may be placed in series with the bleeder resistor and the load applied across the zener. Fig. 58 shows such a circuit. The power supply produces 24 volts of filtered DC, and the bleeder circuit is so designed that nine volts will drop across the zener diode and fifteen volts across the resistor. The load, in parallel with the zener diode, will naturally have nine volts across it, also. When the load current increases, it tries to increase the voltage drop across the bleeder resistor; but since the zener voltage remains constant, its current must decrease. The load current and that through the zener diode will always be exactly the amount required to maintain a constant voltage drop across the bleeder resistor.

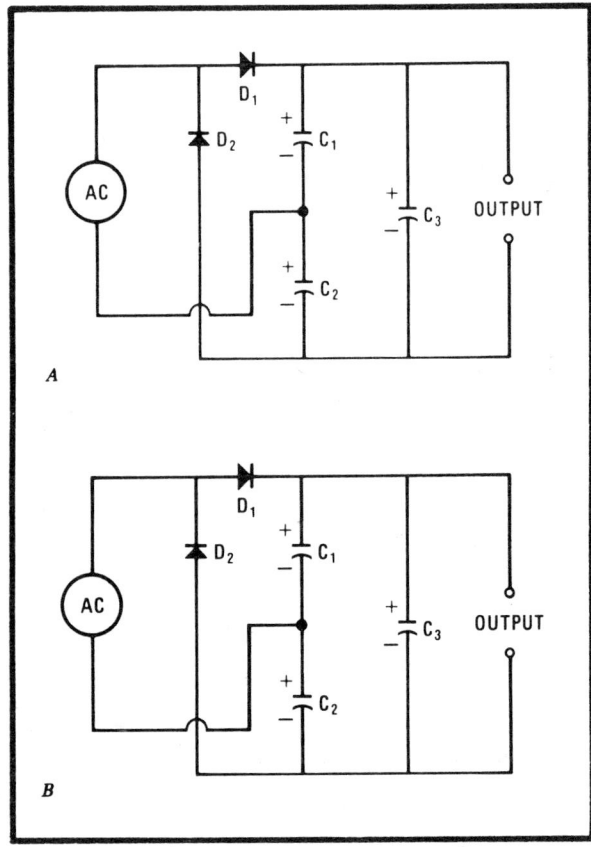

Fig. 57 Full wave voltage doubler
A - In this half cycle, capacitor C_1 charges to full line voltage.
B - During the next half cycle, C_2 charges to full line voltage and since C_1 and C_2 are both in parallel with C_3, the output will be twice the line voltage.

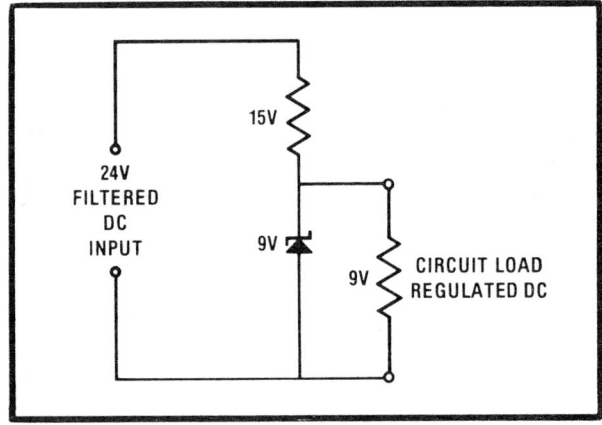

Fig. 58 Zener diode regulated power supply. The zener diode will maintain a constant voltage drop across it by varying its current as the load current fluctuates.

QUESTION:

35. What does an electronic voltage regulator use to sense the voltage being regulated?

SECTION IV:

Logic Circuits

Until the advent of the transistor and the integrated circuit, almost all electronic information was of the analog type. This means that when sound information such as music or voice is to be electronically manipulated, it is first converted into an electrical analogy of amplitude or frequency. This is transmitted to the user, where the variations are once more converted, this time into magnetic variations which move speaker cones and vibrate the air, producing sound waves that can be heard. Vacuum tubes and transistors are able to manipulate these voltage or current variations with little problem and, as the state of the art of electronics developed, it became apparent that the solid state diode and transistors could do more than *vary* the currents and voltages: they could switch a circuit on and off with such speed that they opened up an entirely new field of information handling, that of digital information processing.

Now, electronic computers—not only those we find in business machines, but those in anti-skid brakes, navigation systems, automatic pilots, and many other circuits aboard a modern airplane—are not of the genius type, but are limited in their responses to only two conditions, those representing either Yes or No. Based on the fact that these new electronic devices can switch with unbelievable speed, and are so extremely tiny that thousands can be held on your thumbnail, a form of logic has been adapted and circuits devised that will process input information and furnish *the* logical output from all that is fed in.

In adapting information to a form that can be used in these logic circuits, we must first understand their limitations. A logic circuit can recognize only two conditions: one in which voltage *is present* at a point, and the other in which, voltage is *not present*. These two conditions could be given many different names—we could call voltage present Yes and no voltage No, or we could call voltage Plus and no voltage, Minus. But the conventional way of looking at these conditions is to consider the presence of a voltage as Logic One and the absence of voltage Logic Zero.

A. BINARY NUMBER SYSTEM

Since we must use numbers in almost any information manipulation, we need a number system that can be handled with only these two conditions, 0 and 1. The system that fits this condition is the binary number system and, fortunately, since we must be able to convert our decimal numbers into the binary system, it can easily be done by electronic methods. Let's assume that we want to convert the number 13 into a binary number so it could be processed by an electronic device. Looking at the chart in Fig. 59, we see that it is composed of a series of numbers in which each number is twice the one to its right. Beneath each number required to get our

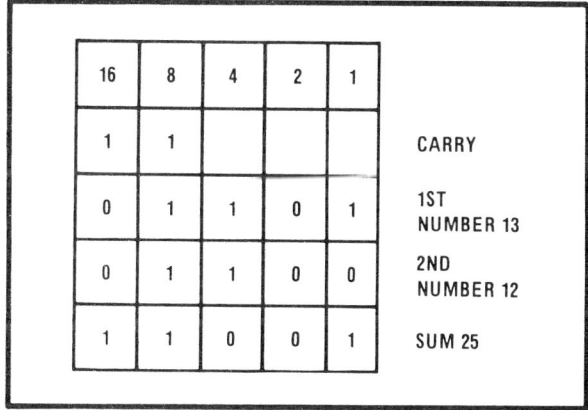

Fig. 59 Binary number addition.

number 13, we put a logic 1; beneath those we do not need, we put logic 0. We do not need a sixteen, but we do need an eight, a four, no two, and a one. (8+4+1=13.) Now if we want to add 12, for instance, to our 13 and do it with binary numbers, we find that for 12 we need no sixteen, one eight, one four, and neither a two nor a one. Now we can add. Start over at the right side where we have a 1 and a 0, which gives us 1. In the two column, we have two zeros, so their sum is zero. In the four column, we have two ones, which gives a sum of one-zero. Now, to handle this, we put down the zero and carry the one over into the eight column. We now have three ones in the eight column, so its sum is one-one, one in the eight column and one to carry over into the sixteen column. Here we have zero-zero-one, which gives us, one. The answer now is 11001; this converts into the decimal system as $16 + 8 + 1 = 25$, which is exactly what we get when we add $12 + 13$.

B. LOGIC GATES

The beauty of this kind of number system is that ultra-tiny, solid state electronic devices can perform billions upon billions of these computations in less time than it takes us to tell about it. These computations, or manipulations, if you want to call them that, are done by the use of gates—simply a fancy name for an arrangement of devices that either conduct or do not conduct.

1. AND gate

One of the simpler gates, and one of the most used, is the AND gate, Fig. 60. In this gate we have two or more inputs, and all of them must have a logic 1 (a voltage) on them in order to get a voltage, or logic 1, at the output.

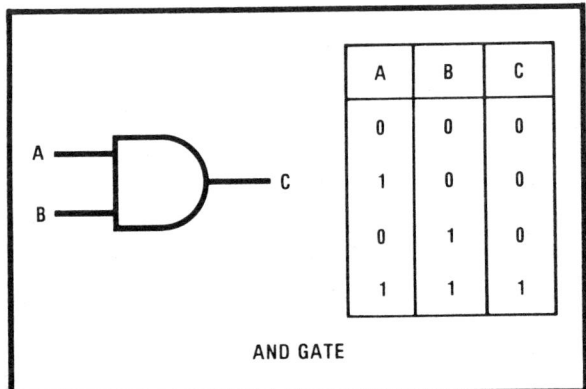

A	B	C
0	0	0
1	0	0
0	1	0
1	1	1

AND GATE

Fig. 60   AND Gate and truth table

We can use a truth table to find out just what the gate is supposed to do, and to check to see that it does it. If we have a logic 0 (no voltage) at both A and B, we will have no voltage, (logic 0) at the output; that's logical. If we have voltage (1) at A, and none (0) at B, we will still have zero at the output. Again, if we have zero at A, and one at B, we will still have zero at the output. But since this is an AND gate, if we have one at A, AND one at B, we will have one at the output. A circuit such as this would be handy for the landing gear lights. All three wheels would have to be up and locked before the gear-up light could illuminate. Any

time any gear is off the uplocks, we have no voltage at that input, and there will be no output.

2. OR gate

Another handy gate is the OR gate. Again, it can have two or more inputs, and any time any one of them has a voltage on it, there will be an output; the truth table shows this. When there is no voltage on A or B, there will be no voltage on C. But if there is voltage on A but not on B, there will be a voltage at the output. Voltage on B and not on A will still produce an output, and if there is a voltage on both A and B, there will be a voltage at the output—since it will work if there is a logic one at either A OR B.

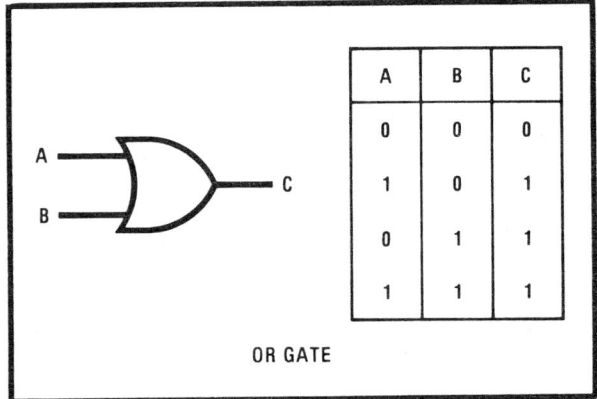

A	B	C
0	0	0
1	0	1
0	1	1
1	1	1

OR GATE

Fig. 61   OR Gate and truth table

This kind of gate circuit would be handy with an electric-pump-type landing gear retraction system, if the up-lock switches have a normally closed position, which is held open by the gear being up and locked; when any of the gear comes off its up-lock, the pump will be turned on until there is no voltage on any input; then the pump will turn off.

3. Amplifier

An amplifier, or buffer, is a circuit with one input

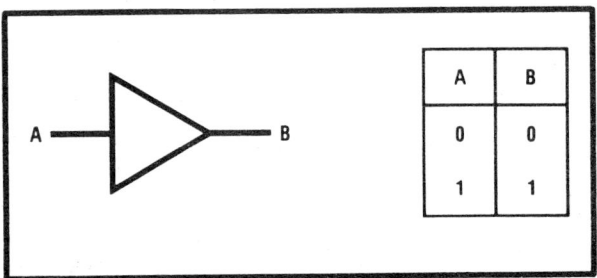

A	B
0	0
1	1

Fig. 62   Amplifier and truth table

and one output, Fig. 62, and has a very simple truth table. If there is no voltage at the input, there will be no voltage at the output. If there is a voltage at the input, there will be a voltage at the output. Circuits of this type can be used to change the values, or to isolate a circuit without changing its logic condition.

4. NOT gate

It is sometimes to our advantage to invert the output of a circuit, and this is done with an inverter. Its symbol is that of an amplifier with a small circle at its output, Fig. 63.

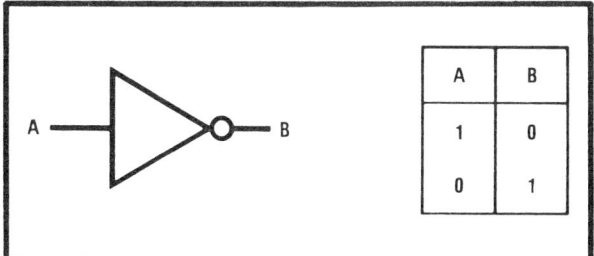

A	B
1	0
0	1

Fig. 63   Inverter and truth table

In this type of circuit, as the truth table shows, the output will be opposite the input. Circuits of this type are used to feed push-pull amplifiers where it is necessary to invert the phase of one of the inputs. An inverted amplifier can be called a NOT gate since, when there is an input voltage, there will NOT be an output voltage.

5. NAND gate

AND,OR, and NOT are the basic circuits or gates, but they have many elaborations, actually combinations, that are useful to the A&P technician. The NAND gate, for example is a NOT

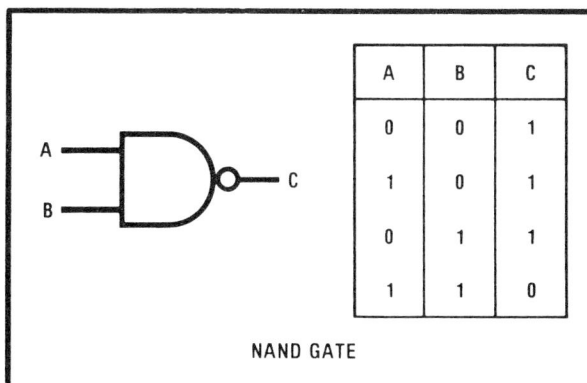

A	B	C
0	0	1
1	0	1
0	1	1
1	1	0

NAND GATE

Fig. 64   NAND Gate and truth table

AND gate and works like an AND gate followed by an inverter. The information it gives is that when both A and B have an input voltage, there will be no voltage at the output, and any time *both* A and B do not have a voltage, there will be a voltage at the output.

6. NOR gate

The NOR, NOT OR, gate acts as an OR gate followed by an inverter. When there is a signal on any of the inputs there will be *no* signal at the output; this is shown in the truth table of Fig. 65.

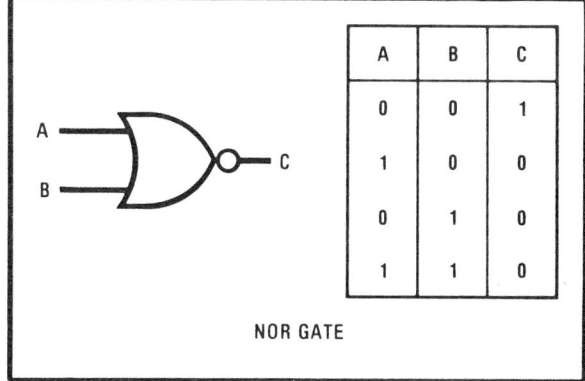

A	B	C
0	0	1
1	0	0
0	1	0
1	1	0

NOR GATE

Fig. 65   NOR Gate and Truth table

7. EXCLUSIVE OR gate

An EXCLUSIVE OR gate has a logic 1, or a voltage, at its output if one, and only one, of its inputs has a voltage present. This is shown by the truth table of Fig. 66.

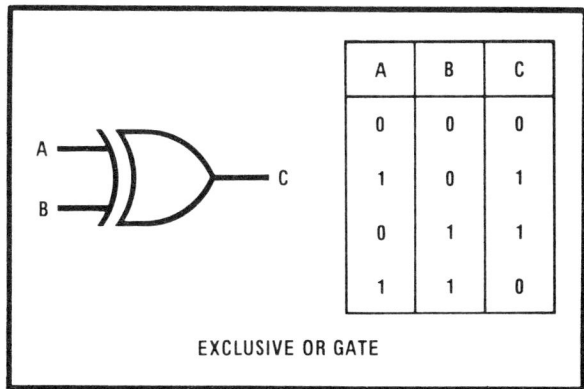

A	B	C
0	0	0
1	0	1
0	1	1
1	1	0

EXCLUSIVE OR GATE

Fig. 66   EXCLUSIVE OR Gate and truth table

By this, you can see that by combining AND, OR, and NOT gates into various arrangements and configurations we can build circuits that can process almost any amount of information and present the logical output and display.

QUESTIONS:

36. What is the decimal equivalent of binary 11010?

37. What would be the binary equivalent of 12?

38. If a two input AND gate has a voltage on both inputs, will there be a voltage at its output?

SECTION: V

Radio Communications

The first really practical application of electronics to aviation was one of communications. When a pilot in an airplane was able to talk with someone on the ground, an entirely new era of utilization of the airplane became possible. Cross-country flying became practical, and navigation without reference to the ground became possible.

A. RADIO WAVES

Earlier we saw that when current flows in a conductor, lines of magnetic force surround the conductor, Fig. 67. If AC flows, this magnetic field starts at the conductor, becomes larger until it reaches its maximum, at the time of the greatest current flow, then drops back to zero and reverses direction as it expands for current in the opposite direction.

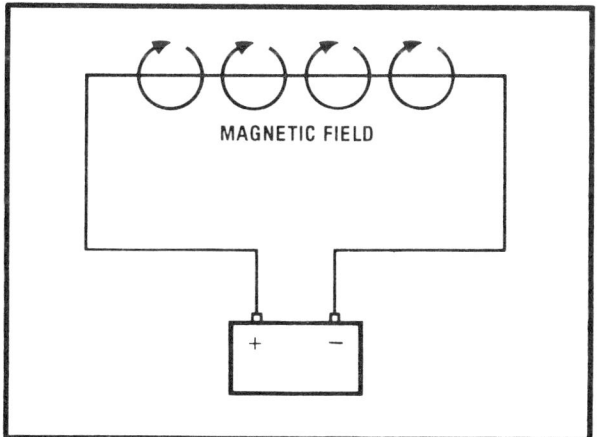

Fig. 67 *A magnetic field surrounds a conductor carrying current. The field is perpendicular to the conductor and its strength is proportional to the amount of current.*

If a source of voltage is connected to a capacitor as seen in Fig. 68, an electrostatic, or electric, field is established between the plates. Now if this capacitor is connected to a source of AC, the electric field will build up, collapse, and reverse as the polarity of the source changes.

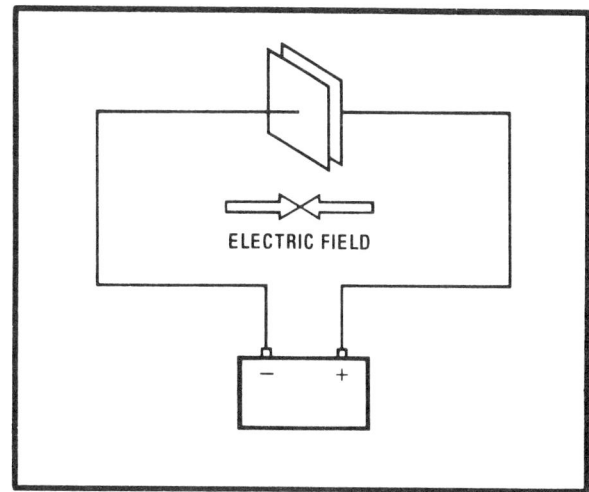

Fig. 68 *An electric or electrostatic field exists when voltage is applied along a conductor or across a capacitor.*

Every conductor possesses some inductance, some capacitance, and some resistance, and there is a particular length of conductor that will be resonant for any given frequency. Electrical energy travels 186,000 miles per second, and the wave length of AC with a frequency of one hundred megahertz, (100,000,000 cycles per second) is 9.82 feet. The higher the frequency, the shorter will be the wave length.

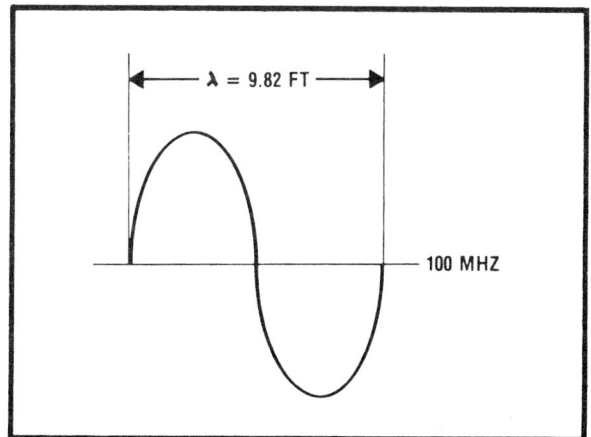

Fig. 69 *Electrical energy travels 186,000 miles per second, [300,000,000 meters per second]. The length of a 100 megahertz wave is three meters, or 9.82 feet.*

If a piece of wire as long as one-half of the wave length of the AC from a generator is suspended in the air, Fig. 70A, and is excited, or fed, in its

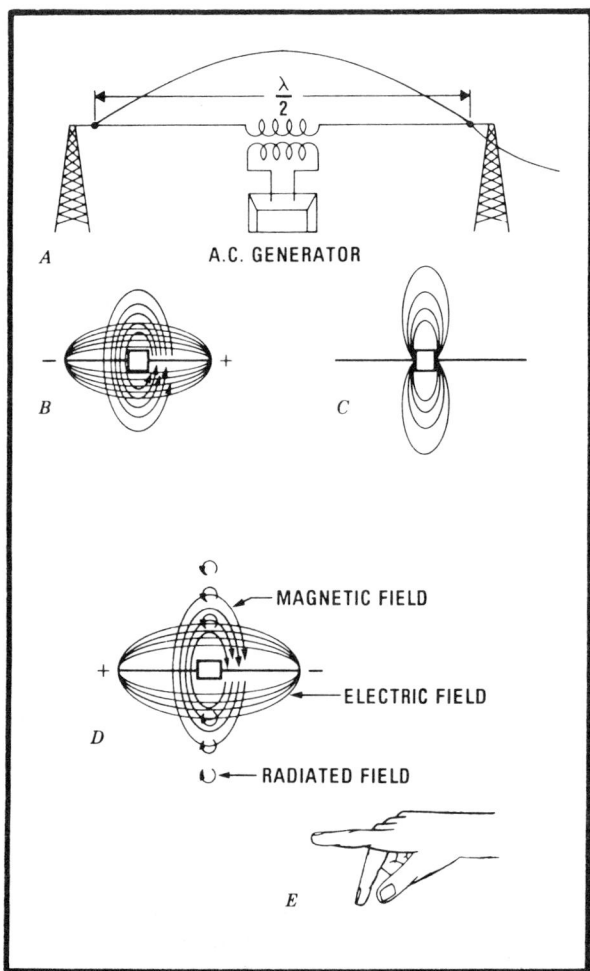

Fig. 70 Generation of radio waves.

A - For maximum radiation, the generator [transmitter] feeds a half-wave length wire in its middle.

B - When the voltage difference between its ends is maximum, the magnetic field has expanded to its greatest dimension.

C - When the high frequency electric field drops to zero to change its polarity, some of the magnetic field is unable to return to the wire.

D - When the electric and magnetic fields again build up, some of the magnetic field that had been unable to return to the wire is radiated out into space.

E - If the fingers of the right hand are arranged in this manner, the forefinger will represent the magnetic field, the thumb, the electric field, and the middle finger will point in the direction of the radiated wave.

middle by the generator, both electromagnetic and electric fields will surround the wire. At the same instant that one end of the wire is negative, the other end will be positive, Fig. 70B, and the electromagnetic field will have expanded to its maximum. As the polarity of the wave starts to change, the magnetic field begins to collapse and the electric field begins to pull in, Fig. 70C. At frequencies of above about ten kilohertz (10,000 cps) the electric field does not have sufficient time to completely pull in before it starts to build up in the opposite direction, and an independent field is formed which is repelled by the buildup of the new electric field. This newly-formed field is then radiated out into space as radio waves, in phase with the fields causing them, but perpendicular to both fields, Fig. 70D. Fig. 70E demonstrates this: if the thumb, forefinger, and middle finger of the right hand are held as shown, with the thumb pointing in the direction of the electric field, the forefinger will point in the direction of the magnetic field, and the middle finger points in the direction of the radiated radio wave, or the direction of propagation.

1. Composition

a. Carrier Waves

A radio wave, to be of any practical use, must of course be able to carry information. In the earliest days of radio communications, waves were sent out—transmitted—in short spurts, with some shorter than others. The shortest spurts were called dots and the longer ones, dashes, and a code was devised so that information could be transmitted as a series of dots and dashes.

Radio communications became even more effective when it became possible to modulate, or change the radio waves to allow them to operate continuously and carry information. The most common form of changing the wave is to vary its strength or to amplitude modulate it (AM).

A radio transmitter has an oscillator that generates AC at a very carefully controlled frequency, assigned by the Federal Communications Commission. This is called the carrier frequency and for aviation communications it is in the very high frequency (VHF) band. For our illustration, let's consider the emergency frequency of 121.5 megahertz (121,500,000 cycles per second). A crystal-controlled oscillator generates this frequency, and it is amplified and fed into an antenna that is resonant at approximately this same frequency, so that a maximum amount of this energy is propagated, or radiated out.

Now, this carrier wave does us little practical good, as it just consists of electric and magnetic fields radiating out into space. In order to use this energy, we must modulate or change it in such a way that the changes will carry the information we want to transmit.

b. Modulation

The signal we wish to transmit, in this case the audio frequency of the voice, is amplified and used to change the amplitude of the carrier, Fig. 71.

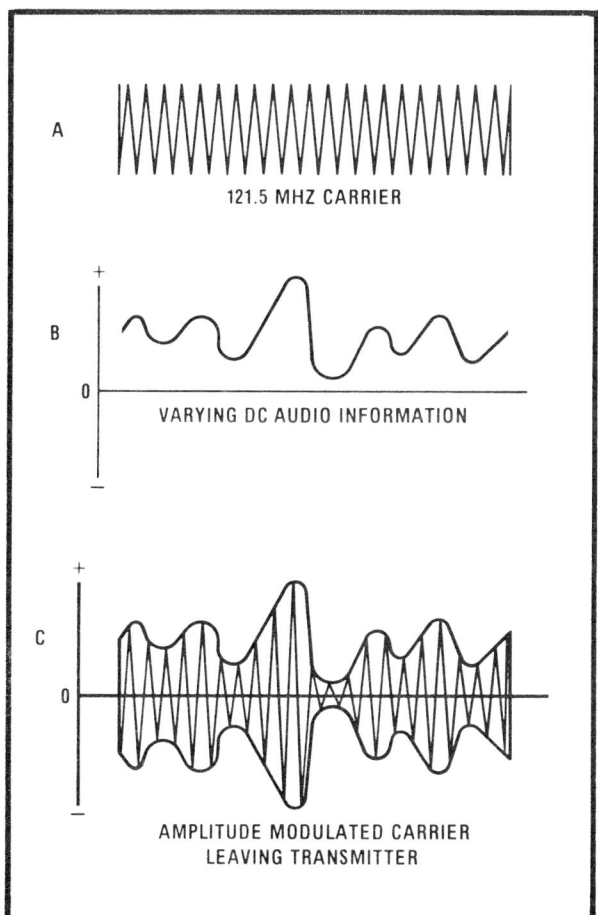

Fig. 71 Composition of a radio wave

A - High frequency AC forms the carrier.
B - DC varying at audio frequencies is developed in the modulator.
C - The carrier is modulated or changed in amplitude to coincide with the audio frequency, and is transmitted in this form.

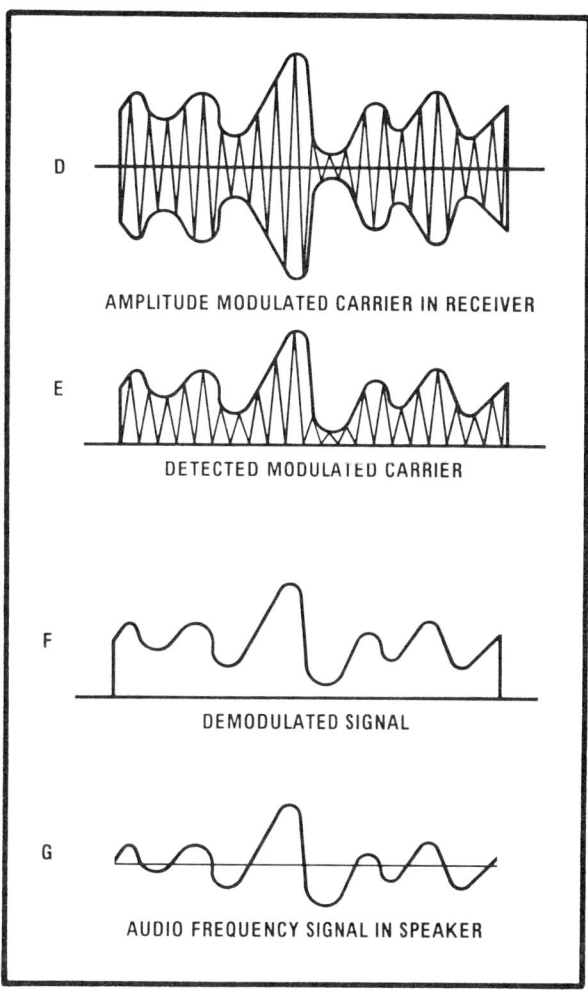

D - The amplitude modulated carrier is received in the receiver in this form.
E - The signal is passed through a detector where one half is removed.
F - A demodulator removes the carrier frequency and leaves only pulsating DC.
G - The pulsating DC is fed into an amplifier where it become AC, just like that produced by the microphone.

Reviewing amplification, we see that the output of an amplifier is varying DC with a waveform determined by the information being amplified. The carrier is AC, and its amplitude, or modulation envelope, conforms to the audio frequency we want to transmit. This modulation is done by superimposing the frequency of the audio on that of the carrier.

It should be shown here what happens to the frequency of the carrier when it is modulated. Because of a phenomenon known as heterodyning, when waves of two different frequencies are mixed there will be two new frequencies generated: one equal to the sum of the two frequencies and the other, the difference between the two. If we modulate a 121.5-MHz carrier with 4000-cycle-per-second audio, the carrier will now occupy a band of frequencies as wide as the carrier and its two sidebands. The lower sideband equals the carrier minus the modulation, or 121.496 MHz, and the upper sideband has a frequency of 121.504 MHz (carrier plus modulation). The bandwidth determines the number of channels that can be spaced in a given frequency band, and as the number of available channels increases, the more critical becomes the carrier frequency, and the less tolerance there is for any frequency drift.

2. Propagation and Reception

When a modulated carrier wave with sufficient energy is fed into an antenna of the proper length, radio waves will be propagated into space. These waves contain both electric and magnetic fields, and when they cut across any conductor in their path, they will induce a voltage in it, a voltage just like the original except of course, much weaker. Much more will be said about the transmitter, the transmitting antenna, and the transmission lines that feed it in a later section; now we are concerned with what happens to the signal after it is picked up by the receiving antenna on the airplane.

B. RADIO RECEIVERS

1. Amplitude Modulation

a. Principle

When the modulated carrier wave cuts across the antenna of a receiver, a voltage is generated in it that is an exact copy of the transmitted signal, but is, as was previously mentioned, much weaker. In order for the information to become usable, it must be modified in the receiver, so it can drive a speaker and be heard.

The most simple form of radio receiver, and one which allows us not only to hear the transmitted signal, but understand the principle of radio reception, is the crystal radio receiver. In Fig. 72 we will develop a crystal receiver and will later apply these principles to a more usable form of aircraft radio receiver.

Radiated energy is picked up by the antenna in Fig. 72A. Not only the signal we want, but every other signal as well. A tank circuit consisting of inductor L_1 and variable capacitor, C_1 in parallel, is connected between the antenna and ground. The value of L_1 and C_1 are adjusted so they will be resonant at the frequency you want to pick up on the receiver.

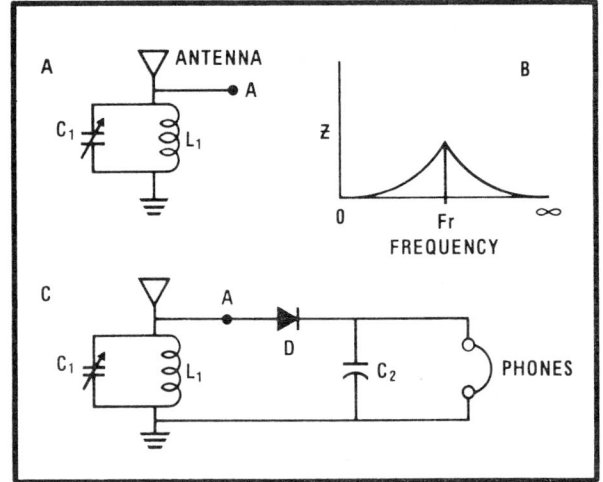

Fig. 72   Principle of a crystal radio receiver
A - The radio signal is received by the antenna and all but the desired frequency is passed to ground through the tank circuit, L_1 - C_1.
B - Frequencies above resonance go to ground through the capacitor, and those below resonance pass to ground through the inductor.
C - The signal is detected with diode D, demodulated with capacitor C_2, and appears as AC in the phones.

We can omit a lot of mathematics involved in resonant circuits by visualizing the change in impedance with frequency as seen in the curve of Fig. 72B. Frequencies below resonance will find a path to ground through the low inductive reactance of coil L_1, and frequencies above resonance will be conducted to ground through the low capacitive reactance of capacitor C_1. The frequency to which the circuit is tuned will produce the maximum voltage on the antenna side of the tank, at point A, Fig. 72A.

If we were to look at the voltage between point A and ground, using an oscilloscope, it would look like the voltage wave of the modulated carrier shown in Fig. 71D. The net voltage is zero because there is as much positive voltage as there is negative.

The selected modulated carrier wave leaves point A, Fig. 72C, and passes through diode D, which acts as a check valve and passes only one-half of the signal. The produces a rectified wave such as Fig. 71E. It is still unusable because it is at the carrier frequency and far too high to be heard by the human ear.

The rectified or detected signal now goes to the capacitor C_2 and the phones, where the carrier is removed or the signal is demodulated. Remember that the modulated carrier consists of two AC signals, one at the carrier frequency and one at the audio frequency. Capacitor C_2 has a very low capacitive reactance at the carrier frequency, but an appreciable reactance to the audio frequency; so the carrier will be passed on to ground through C_2, and a varying DC similar to Fig. 71F will be delivered to the phones.

Earphones are made of a coil of wire wrapped around a small permanent magnet; this coil carries the demodulated signal which is varying DC. The magnet is placed near a thin, spring steel diaphragm and holds it partially deflected toward the magnet, Fig. 73. When the varying voltage through the coil increases, the electromagnet formed by the coil aids the permanent magnet and pulls the diaphragm farther over, but when the voltage decreases, the strength of the combined magnets decreases, and the diaphragm is allowed to relax. The movement of the diaphragm is in step with the audio voltage which has been taken from the carrier, and its movement produces sound waves which affect the ear in exactly the same way as the original sound back at the transmitter.

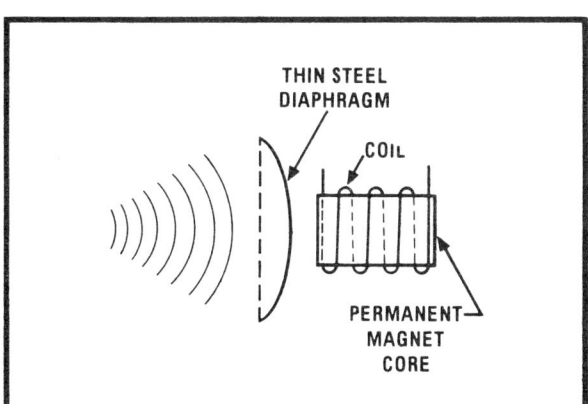

Fig. 73 *Earphones convert the AC fed them into pulsations of the thin steel diaphragm which produces sound waves just like the ones which produced the signal in the microphone.*

b. Superheterodyne Receivers

The simple receiver just described will allow you to hear a very strong radio signal, but it gives you very little output as there is no power supplied to it, nor does it give you very much choice in the stations you can receive.

The superheterodyne is one of the most popular receiver circuits in current use, as it allows its various stages of amplification of be very selective and concentrates its energy in the production of the one signal desired.

In Fig. 74, a tuned pre-amplifier selects a rather narrow band of frequencies picked up by the antenna, in a manner similar to that previously described, and amplifies them, by allowing them to control a flow of electrons from the power supply of the receiver.

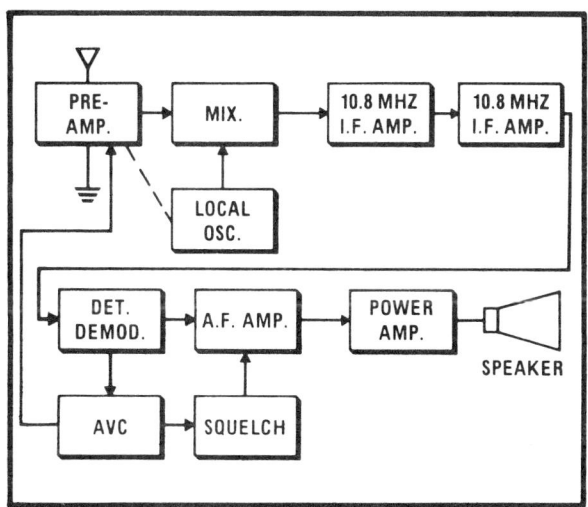

Fig. 74 *Block diagram of a superheterodyne receiver.*

The amplified, modulated carrier is now directed into the mixer stage, sometimes called a converter. Also fed into this mixer is the output of a local oscillator. The frequency of the local oscillator is controlled by the tuning of the pre-amplifier so it will always oscillate at a frequency a specific number of cycles away from that of the RF energy amplified by the pre-amplifier. In the example of this simple superheterodyne block diagram we will assume that the pre-amplifier is tuned to 121.5 MHz, and when selecting this, the local oscillator was tuned so that it produced a frequency of 110.7 MHz. Both of these signals are fed into the mixer where the process of heterodyning, as previously mentioned, causes the formation of two new frequencies, the sum of the two, 232.2 MHz, and the difference between them, 10.8 MHz. All four frequencies are fed from the mixer into the intermediate frequency (IF) amplifier which is highly selective and amplifies only the difference frequency, 10.8 MHz. All of the other frequencies are passed on to ground in this stage. When another frequency is selected on the receiver, for instance 122.9 MHz, the local oscillator will automatically produce 112.1 MHz. This time the IF amplifier will receive 112.1, 122.9, 235.0, and 10.8 MHz. Since the IF amplifier is highly selective, it will amplify only the 10.8 MHz frequency and pass it to the next stage; all of the others are passed to ground.

The intermediate frequency is modulated and is exactly like the original received signal, except for its frequency. The reason for going to this intermediate step is to get the maximum efficiency from the equipment. The pre-amplifier must be rather broad-banded in its characteristics since as part of an aircraft communications receiver must amplify all of the frequencies in the range of 118.0 to 135.95 MHz, and it cannot do this with optimum efficiency over the entire range. By converting these frequencies into one—10.8 MHz, in this example—the stages of intermediate frequency amplification have only one frequency to amplify and they can be made highly efficient.

The second stage of intermediate frequency amplification further amplifies the signal and delivers it to the detector-demodulator stage where the modulated intermediate frequency signal is detected, or rectified, making it into varying DC, but still having the 10.8 MHz component. The demodulator removes this and delivers to the audio frequency amplifier a changing DC with no component of either the original carrier or the intermediate frequency. The AF amplifier then builds up the voltage of the audio signal and delivers it to a power amplifier where sufficient current is controlled to drive speakers with any required volume.

The automatic volume control (AVC) circuit works to maintain the audio output at a rather constant value. If the input signal is strong and there is a tendency for the output to get too loud, a signal from the detector-demodulator stage is sent back to the original RF amplification stage; its voltage

output, or gain, is cut down so it will produce the desired audio output.

Aircraft communications receivers have such a high gain that they amplify the internal noises in the receiver, and when they are tuned to a channel with no signal coming over it, there will be a hissing noise that is distracting, but when a signal comes in and is amplified, it will override the hissing noise. To minimize this distraction, a squelch circuit is installed. The same voltage that drives the automatic volume control is applied to the audio amplifier so that when there is no signal being received, the audio will be down low enough that no hissing will be heard, but when a signal is received, the squelch will allow the audio frequency amplifier to amplify it to the proper volume.

QUESTIONS

39. Which antenna would be the longer, one designed for UHF, or one for VHF?

40. What changes when a carrier is amplitude modulated?

41. What is meant by heterodyning?

42. What does the IF amplifier amplify in a superheterodyne receiver?

43. What is the purpose of an AVC circuit in a superheterodyne receiver?

C. Double Conversion Superheterodyne Receivers

The simple superheterodyne receiver just described has all of the basic principles that makes for good communications, except that the crowding of the available frequency channels requires more precise control of the frequencies and closer spacing of adjacent channels. In the days when there were three, five, or maybe even ten channels in use, each channel could have its own crystal to control the frequency, but now with 360 and even 720 channels in the same frequency band, closer control of frequency is required, and crystals must generate more than one frequency in both the receiver and the transmitter. In the receiver we use double conversion, meaning that we have two local oscillators, one controlled by the megahertz selector switch on the face of the receiver and the other controlled by the kilohertz selector.

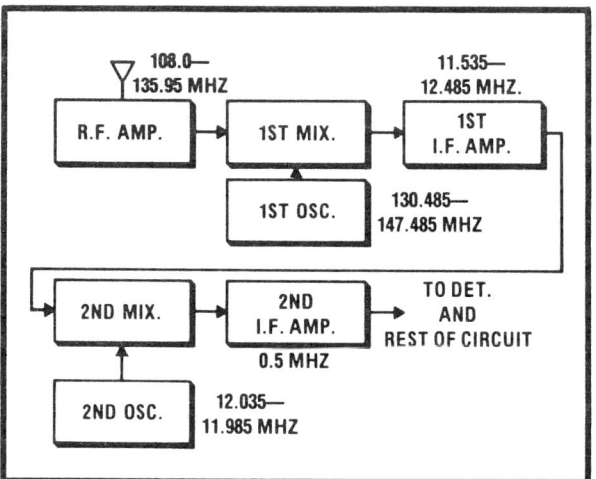

Fig. 75 Block diagram of a double conversion superheterodyne receiver.

The first intermediate frequency amplifier amplifies and passes a band of frequencies rather than one discreet frequency, as the simple superheterodyne did. The output of the second oscillator heterodynes, or beats, with this first intermediate frequency to produce a second IF, which is taken through the rest of the circuit as was done in the simple circuit.

2. Frequency Modulation

a. Principle

One of the primary limitations of amplitude modulation is its susceptibility to static. Static, or atmospheric noise, is essentially an amplitude variation, and the reception of intelligible information can be improved by using a system of modulation that does not use amplitude variations.

In a frequency-modulated system, AC is generated in the very high frequency band, (VHF) for the carrier, and this is modulated with the information to be transmitted, so that it changes from its quiescent, or center frequency by an amount proportional to the amplitude of the modulation. For example, if the transmitter has a center frequency of 112.0 megahertz and it is modulated with an audio frequency of, say, 1000 hertz. The carrier will vary from its center frequency by an amount proportional to the

amplitude of the 1000 hertz signal, and it will vary the amplitude one thousand times a second. This varying, or swinging of the carrier from one side of center to the other, generates sidebands similar to those in amplitude modulation. These sidebands, however, instead of having to be added to the carrier power, are produced by the carrier; so the antenna current from an FM transmitter does not vary appreciably between a signal and no-signal condition. Modulation of an FM signal also requires much less power, because it is the oscillator that is modulated rather than the power amplifier as must be done with AM.

b. FM Receivers

An FM receiver, Fig. 76 uses a superheterodyne circuit and has two stages that differ from those in An AM superheterodyne receiver. The signal is received by the antenna and taken into the tuned RF amplifier.

The local oscillator is tuned with the RF amplifier so that the proper intermediate frequency will be generated. This intermediate frequency is amplified through a couple of stages of IF amplification and then it passes to a limiter. The signal, up to this point, varies in amplitude because of noise and static, but in the limiter stages it is clipped in both polarities, so the signal fed to the discriminator varies in frequency but not in amplitude.

The discriminator changes the frequency variations into amplitude variations, and the signal sent to the audio frequency amplifiers is the same as they would receive in an AM receiver.

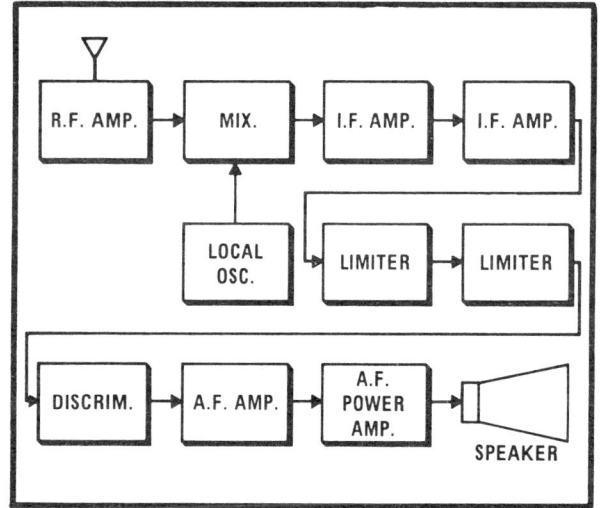

Fig. 76 Block diagram of a frequency modulated [FM] receiver.

c. FM Uses

In the commercial broadcast field, FM is used because it has the capability of transmitting a much wider band of frequencies than amplitude modulated transmitters. Some land mobile communications is done with frequency modulation, but the prinicple use of FM in aviation is the VOR navigation signal. As will be explained in the appropriate section of this text, the VOR signal consists of two components; one amplitude modulates the transmitter and the other frequency-modulates it. The FM portion of the signal is at a much higher frequency than the AM portion, and they are separated by the use of filter circuits and handled as separate signals.

C. RADIO TRANSMITTERS

Now that we understand what a radio wave is, and what happens to it when it gets into the receiver, let's back up just a little and see how it got there in the first place.

Regardless of whether we are talking about a broadcast transmitter, the CB radio in your car, or the communications transmitter in an airplane, the principle is the same. We must have an oscillator that generates an AC of the frequency assigned for the particular transmission; we must have amplifiers that build this AC up to the value needed for transmissions; and we must have a modulation system that changes the sound energy in our voice into electrical signals and modulates or changes the carrier wave so it will have the voice signal on it. There must also be an antenna that will radiate this energy into space, and finally there must be the necessary transmission lines to carry this modulated carrier wave to the antenna, with the minimum of losses.

The spectrum of radio frequency has been divided up by international treaties into various bands and ranges, and certain of them have been assigned for aeronautical communications. Table 1 shows the basic frequency divisions, with their names and the aeronautical radio facilities in each band.

The vast majority of aeronautical communications is done in the very high frequency (VHF) band, between 108.0 and 135.95 megahertz, and it is with this type of equipment that we will be most concerned in this treatise.

VHF communications is, because of the nature of its waves, restricted to line-of-sight distances.

FREQUENCY BAND	UTILIZATION
Low Frequency Below 300 KHz	LF/MF Communications - 200 - 416 KHz ADF Ranges - 90 - 1800 KHz Loran 1750 - 1950 KHz
Medium Frequency 300 - 3000 KHz	Commerical AM Broadcast 535 - 1605 KHz
High Frequency 3 - 30 MHz	HF Communications 2 - 25 MHz
Very High Frequency 30 - 300 MHz	Marker Beacons 75 MHz VHF Navigation and Communications 108.0 - 135.95 MHz
Ultra High Frequency 300 - 3000 MHz	Transponder and DME 960 - 1215 MHz
Super High Frequency 3000 MHz - Up	Radar 220 - 5200 MHz

Table 1 Frequency Utilization

This means you can receive a signal from a transmitter only if there are no obstructions between the two antennas. Because of the curvature of the earth, the maximum range of line-of-sight communications is restricted. Maximum reception range (in miles) is about 1.41 times the square root of the altitude (in feet) of the receiver above the transmitter. For example, an airplane flying at 10,000 feet above the ground would have a maximim reception distance of about 141 miles. Of course there are many things that could decrease this distance.

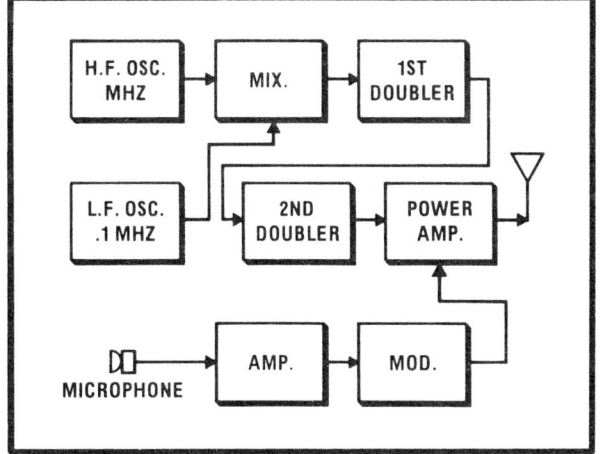

Fig. 77 Block diagram of a VHF Transmitter

Fig. 77 is a block diagram of a VHF transmitter. The two oscillators are crystal controlled and produce a stable frequency for the carrier. The high-frequency oscillator has its crystals selected by the megahertz selector and the low-frequency oscillator by the kilohertz selector. The output of both oscillators is fed into a mixer and then filtered to get the desired frequency. In the VHF range, the carrier frequencies are higher than it is possible for crystals to oscillate, so the output of the mixer is fed into a circuit which doubles the frequency that comes from the mixer, and then into another doubler. The frequency that reaches the power amplifier is, therefore, four times that which came from the mixer.

The sound waves that enter the microphone are converted into electrical signals and amplified, then taken into a modulator which works with the power amplifier to modulate or superimpose the audio signal on the carrier.

QUESTIONS

44. What is the purpose of a double conversion superheterodyne receiver?

45. What is changed when a carrier is frequency modulated?

46. Why does an FM signal require less power for transmission than an AM signal?

47. What are four basic requirements for a radio transmitter?

48. What is one limitation of VHF communications for aircraft use?

D. ANTENNA

1. Principle

The modulated carrier wave generated in the transmitter is carried to the antenna by a special wire called a transmission line—about which we will say more. Now we wish to consider the antenna itself:

As we saw in Fig. 70, the signal that enters the antenna consists of both electric and magnetic fields, and if it is fed into a wire exactly one-half of the wave length of the carrier, the wire will act as a resonant circuit, Fig. 78. Electrons will flow to one end of the antenna and pile up there; then, during the next half cycle, they will all travel to the opposite end where they again pile up. When all of them are at one end, the antenna will have no current flow, and the magnetic fields will be zero. The electric fields will be maximum, however, because of the voltage difference between the two ends. As the electrons start back

to the other end, the electric field decreases and the magnetic field increases, until at the point at which the ends are at the same voltage and the current is maximum. As long as the signal is supplied to the antenna, it will oscillate, and since it is doing this at radio frequencies, the field expands so far each cycle that part of it cannot get back into the antenna before the next cycle forces it back out, and part of the energy is radiated into space.

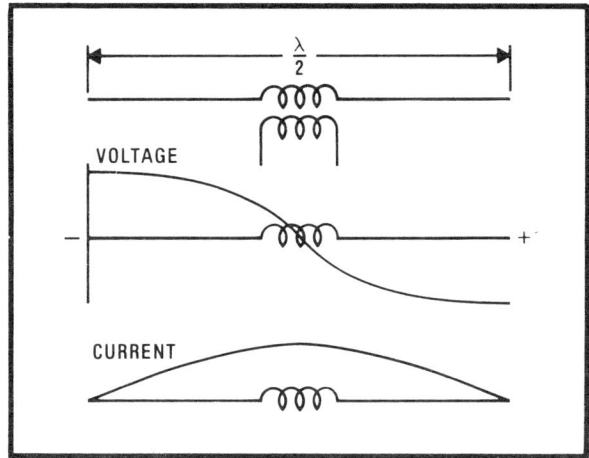

Fig. 78   The half-wave antenna
A - For maximum radiation, an antenna should be one half of a wave length long. This one is center-fed.
B - At the instant the voltage is maximum at its ends,
C - The current will be maximum at its center.

2. Length

As we can see, the greatest amount of energy will be radiated into space if the antenna length is exactly one-half wave length. If the transmitter is used for more than one frequency, there must be either some means of adjusting its length or else the antenna must be a compromise and not operate at its peak efficiency. Since the antenna acts as a resonant circuit, and resonance may be controlled with either capacitance or inductance, adding either of these in series with the antenna will change its resonant frequency and effectively change its electrical length. If a capacitor is added in series with the feed line, the antenna will be electrically shorter, and if an inductor is added, it will be electrically longer.

The formula for finding the required length for a half-wave antenna is:

$$\text{Length (feet)} = \frac{468}{F \text{ MHz}}$$

The constant 468 comes from the fact that radio waves travel at 300 million meters per second, and therefore the wave length of a one megahertz signal is 300 meters. One-half wave is 150 meters, or 492 feet.

The dielectric effect of the air at the end of a half-wave antenna effectively lengthens it by about 5 percent, so the antenna should be computed at 95% of 492, or 468 feet. This would be correct length for one megahertz, and to find the length for any other frequency, divide 468 by the frequency in megahertz.

3. Polarization and Field Pattern

If the antenna is horizontal to the earth, the energy will be radiated out from the antenna in a doughnut pattern, Fig. 79-A, and there will be no lengthwise radiation from it. The maximum signal will be radiated perpendicular to the conductor. Fig. 79-B is a polar diagram of a horizontal antenna, showing the way the relative strength of the signal varies as the direction from the antenna changes.

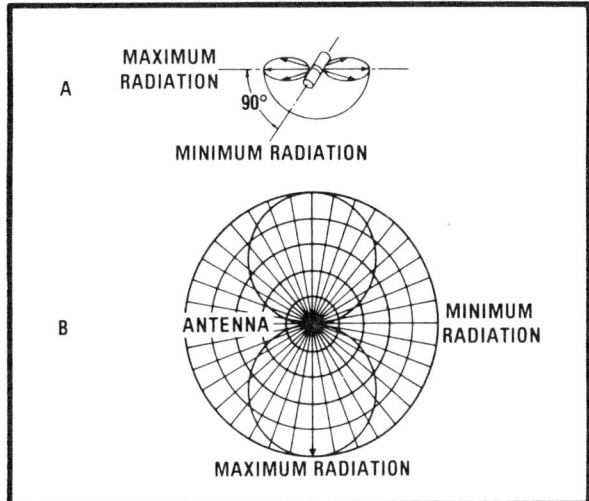

Fig. 79 A horizontally polarized antenna radiates the maximum amount of energy perpendicular to its length.

Maximum reception will be obtained from a horizontally oriented transmitter antenna if the receiver antenna also is horizontal.

If a transmitter antenna is mounted vertically, it will still put out a field pattern resembling a doughnut, but this time it will be as though the doughnut were laid on the ground, Fig. 80-A, radiating energy equally in all directions, Fig.

80-B. Any signal transmitted from a vertical antenna will best be received on a vertical antenna.

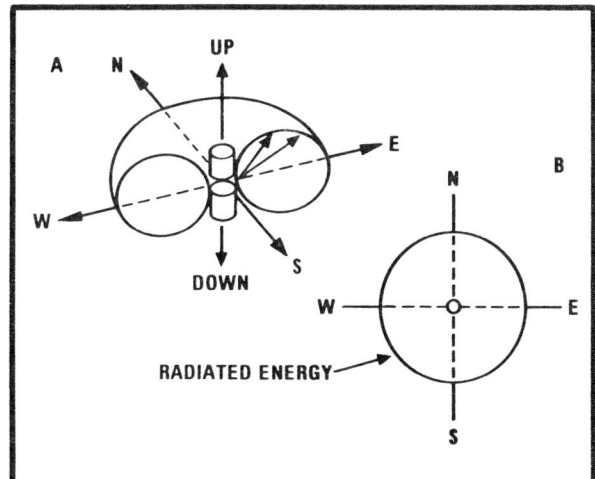

Fig. 80 A vertically polarized antenna radiates its signal equally strong in all directions.

4. Types

a. Hertz

The half-wave antenna we have just been discussing is called a Hertz antenna and may be fed either in the center, as we have shown, or from the end. If it is fed in the center where the current is maximum, it is called a current-fed antenna, but if it is fed at its end where the current is zero and the voltage is maximum, it is said to be voltage-fed.

A form of Hertz antenna is the dipole, which is a half-wave antenna consisting of two quarter-wave arms, fed in the center. The VOR antenna found on most all aircraft is a special form of dipole. It is horizontally oriented, and the arms form a V, with its apex in the line of flight, usually facing forward.

b. Marconi

It is not always convenient to use a half-wave antenna on an aircraft, so a quarter-wave antenna is often used. With this type of antenna, all of the requirements for transmission can be met with a quarter-wave conductor by using a reflector to serve for the other quarter wave. Aircraft communications antennas are usually of the quarter-wave length, vertically-polarized type, and because of the wide range of frequencies handled, are usually a compromise with regard to length. The skin of the aircraft serves as the reflector, or ground plane, Fig. 81-A. Sometimes VHF stations on the ground mount their antenna on the top of a pole, and since there is no structure to serve as a reflector, four or more wires, Fig. 81-B, each as long as the antenna, are used to form the ground plane.

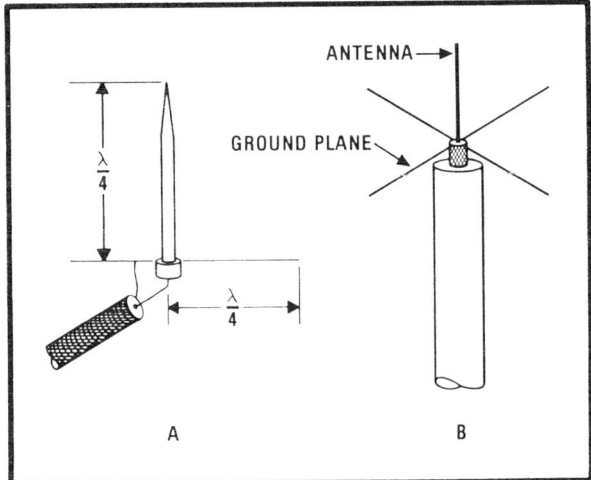

Fig. 81 Quarter-wave vertically polarized antenna.

A - Airborne antennas use the skin of the aircraft for the ground plane.

B - Quarter-wave antenna mounted on poles require quarter-wave radial arms to serve as the ground plane.

When a quarter-wave antenna is mounted on a fabric-covered airplane, it is wise to put a sheet of aluminum foil large enough to serve as the ground plane, inside the airplane, under the fabric.

c. Loop

As we have seen, a vertically-polarized antenna is omnidirectional; that is, it receives a signal equally well from any direction, Fig. 80, and a horizontally polarized antenna will receive a signal best from a station at either side, perpendicular to the antenna. Its strength will be the least for a signal from a transmitter in line with its length, Fig. 79.

The directional characteristics of an antenna may be enhanced by winding it in the form of a loop, Fig. 82-A. A signal received from a transmitter in line with the plane of the loop, Fig. 82-C, will be received on side *A* before it is received on side *B*, and the voltage induced in side *A* will be out of

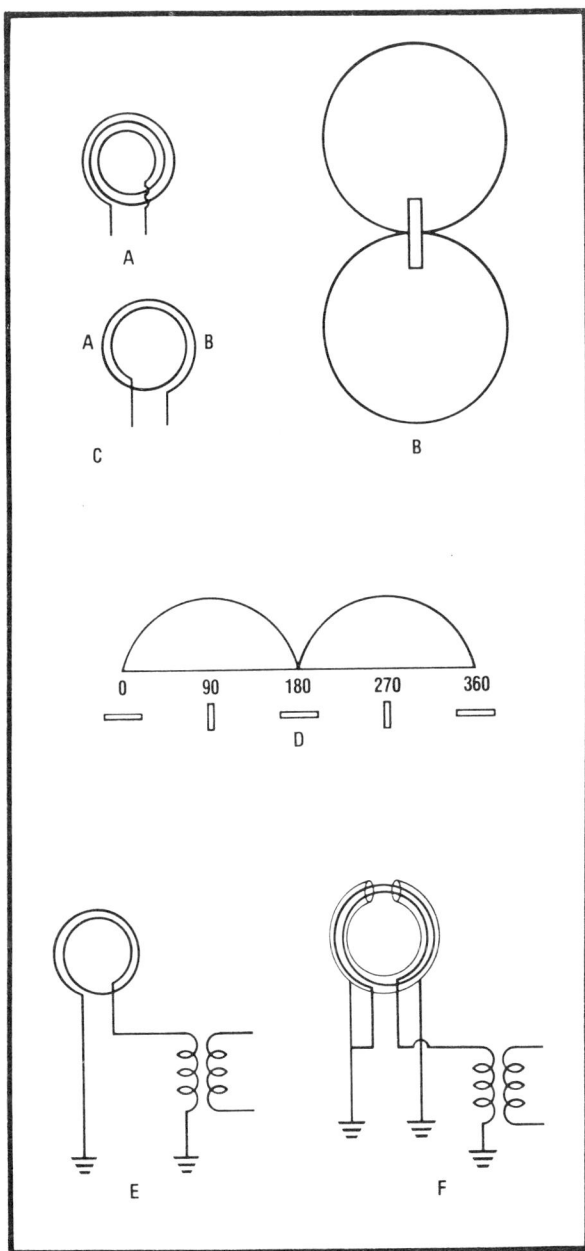

Fig. 82 Loop antenna

A - An antenna wound in the form of a loop is highly direction sensitive.

B - Maximum signal strength is received in line with the antenna, but sharp nulls exist on either side, broadside to the plane of the loop.

C - The signal received in side A will be out of phase with that received in side B.

D - Rotation of the loop produces a sharp null as the plane of the loop is broadside to the station.

E - A loop with one end grounded produces an unbalanced load, and has low efficiency.

F - A shielded loop is more efficient than an unshielded one.

phase with that induced into side *B*. The antenna current will then be the difference between the current in the two sides.

When a signal is received from a transmitter directly broadside to the loop, the voltages induced in the two sides will be equal to each other, but opposite in polarity and will cancel. This characteristic makes the loop antenna useful for direction finding.

In the early days of radio navigation, a loop was strung inside the fuselage behind the cabin and when the airplane was flown on the heading which produced the weakest signal, it was pointed either directly toward the station or directly away from it. The next step in radio navigation was to use a loop which was rotatable from the cockpit, and an azimuth scale was connected to the loop. A station was tuned in and the loop was rotated until the signal strength became the weakest, indicating that the loop was pointing directly broadside to the transmitter. By reading the azimuth scale, the pilot could find the relative bearing of the transmission to the nose of the airplane. One of the major problems was that the loop had no way of determining on which side the station was located. This problem was called 180° ambiguity.

A quick glance at Fig. 82-D shows the reason for tuning to a null rather than to a peak. The null is much sharper, meaning that a small rotation of the loop here will produce a much greater volume change than it would at a peak.

A loop antenna with one end grounded and the other end feeding the receiver produces an unbalanced load, and its efficiency is low. In order to overcome the effect of the unbalanced loop, a metal shield with a break in its middle may be used to encase the loop. The wave from the transmitter can penetrate the shield and induce a voltage in the loop. Current will also be induced into the shield, but this current splits and goes to ground. The shield current in flowing to ground induces voltages into the coil, but since these voltages are of opposite polarity, they cancel each other.

The loop antenna, as we mentioned, was first used as a fixed loop for heading information, then as a rotatable loop for radio direction finding (RDF), and finally as a servo-driven loop for automatic direction finding (ADF)—which we will cover in more detail later. The rotatable loop used for

modern ADF systems is wound on a powdered iron core, that concentrates the signal from the radio transmitter, and allows the use of a much smaller loop than is required when using an air core.

5. Voltage Standing Wave Ratio [VSWR]

Since an antenna is a resonant circuit, the voltage will vary up and down in its length. When it has exactly the correct length and is properly fed, the voltage will be maximum at its ends and zero in the center; but if the antenna does not provide an exact match for the transmission line, some of the energy will be reflected back from the antenna and will represent a loss. A perfect match would give a VSWR of 1.0, and the higher the standing wave ratio, the less efficient the antenna. VSWR will differ for every frequency, and ideally the antenna should have a flat VSWR curve, as near 1.0 as possible. Fig. 83 shows a VSWR vs Frequency curve of a very good communications antenna. Bent-wire antennas, such as are used for some of the lower cost installations, have a standing wave ratio as high as six or even seven.

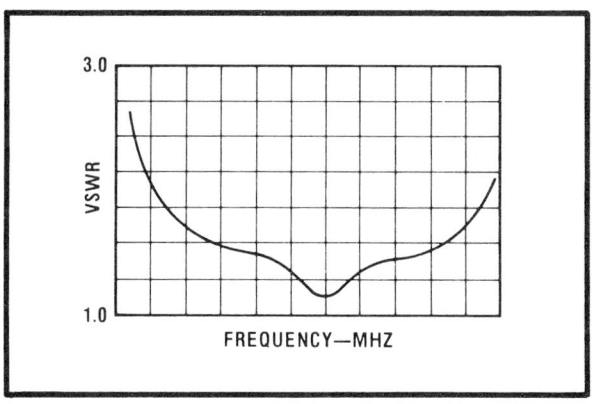

Fig. 83 The Voltage Standing Wave Ratio of an antenna varies with the frequency. Ideally, the VSWR should be as near 1.0 as possible.

QUESTIONS

49. What portion of a wave length should an antenna be for the maximum radiation of energy?

50. What is the correct length for a half-wave antenna for 75 MHz?

51. What is the purpose of the ground plane used with a quarter-wave antenna?

52. Does a loop antenna receive a signal with the greatest intensity when it is pointed toward the station or when it is broadside to it?

E. TRANSMISSION LINES

If we consider the antenna to be the electrical load and the transmitter to be the generator, in order to get the maximum amount of power to the antenna, the load and generator must be matched; that is, the impedance of the load should equal the impedance of the source. Aircraft antennas are considered to have a nominal impedance of 50 ohms, and this should be matched with a transmission line having as near the same impedance as possible. Almost all aircraft antennas are connected to the receiver or transmitters with coaxial cable, which has a nominal impedance of about 53 ohms.

RG-58A/U coaxial cable is the most commonly used transmission line for aircraft radio equipment and consists of a center conductor, a polyethylene dielectric, and a copper braid woven over the dielectric to form the outer conductor. Over this braid is a waterproof vinyl covering. Coax, as this type of cable is called, has the advantage that it has no external field, and it is not susceptible to external fields.

Most RG-58A/U coax is terminated with a form of connector called a BNC connector, and this attaches to the equipment with a quarter-turn twist which provides a good tight seal for both the center conductor and the outer braid. Fig. 84 shows typical terminations of coax cable with BNC connectors.

Coaxial cable is an unbalanced transmission line, and in order to match it to a balance antenna such as the dipole used for a VOR receiver, a special matching transformer must be used. Fig. 85 shows the most simple form of VOR antenna. A phenolic block mounts on the vertical fin of the airplane, and antenna rods stick out from the block to form a V. If the transmission line were attached directly to the rods, with the inner conductor to one rod and the outer conductor to the other, an unbalanced field would result.

To prevent this, a balun is used. A balun may be thought of as a transformer that matches *balanced* antenna to an *un*balanced line. The simple balun used with this type of antenna is similar to the detail of Fig. 85-B.

A piece of transmission line has its outer conductor cut, spread back, and terminal lugs attached to it, so it may be connected to the

antenna rods. The center conductor is left open and the outer conductor is grounded to the airframe about a quarter-wave-length back from the point of attachment to the antenna rods.

You can see from Fig. 85-A that each antenna rod feeds a piece of the outer conductor of the coax that is a quarter-wave-length long, and this signal is induced into the center conductor. The more elaborate VOR antennas have the balancing network installed in the antenna themselves.

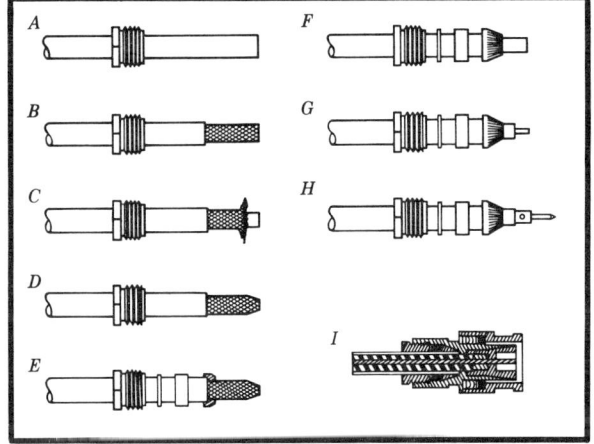

Fig. 84   Termination of Coaxial cable transmission line:

A - Slip the nut over the end of the cable.

B - Remove about one half inch of the outer jacket, being careful that you do not nick the outer conductor.

C - Push the outer braid back so you can cut off about one eighth of an inch of the inner conductor and the insulator.

D - Taper the braid.

E - Slip the washer, gasket and sleeve over the outer braid so the sleeve fits square against the end of the outer jacket.

F - Comb the braid out smooth and fold it back over the sleeve, then trim the ends as shown.

G - Cut back the inner insulator to expose about one eighth inch of the inner conductor. Be careful that you do nick it.

H - Tin the center conductor and solder the tip in place. Be careful that the dielectric is not overheated.

I - Slide the body over the tip, and back as far as it will go. Screw the nut into the body and tighten it.

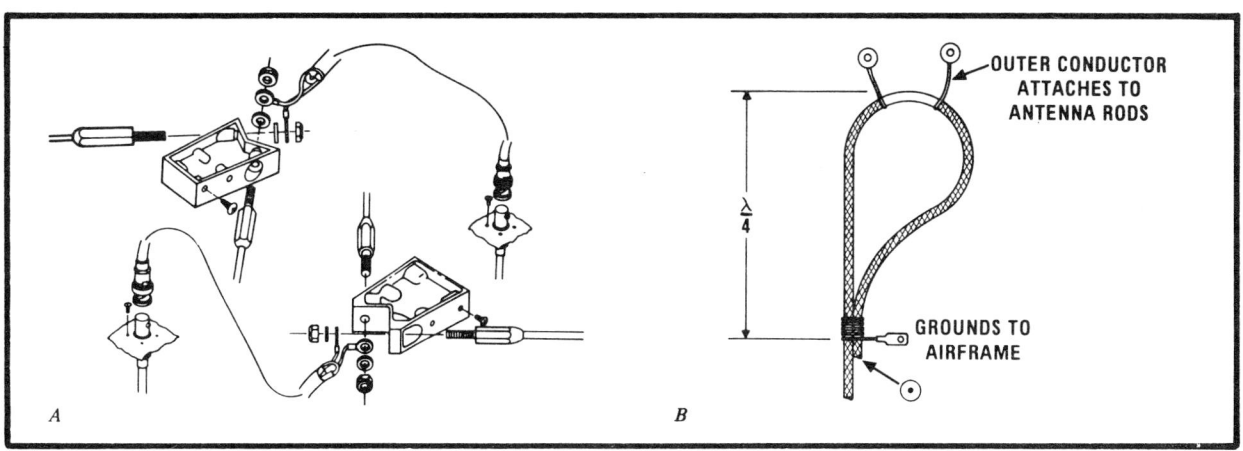

Fig. 85   Simple VOR antenna
A - The most simple VOR antenna consists of two rods extending from a phenolic block.

B -A balun is used to match the unbalanced coaxial transmission line to the balanced dipole antenna.

OUTER CONDUCTOR ATTACHES TO ANTENNA RODS

$\frac{\lambda}{4}$

GROUNDS TO AIRFRAME

QUESTIONS

53. What is one of the main advantages of using coaxial cable for aircraft transmission lines?

54. What is the purpose of the balun connected to a VOR antenna?

SECTION: VI
Radio Navigation

The airplane could be really useful as a means of transportation only when it became possible to fly from point A to point B under conditions when the ground below was not visible. In order to bring this about, some means of establishing an electronic path through the sky was needed, and radio navigation came into being.

Early in the history of radio navigation, the four-leg, low-frequency radio range was developed. In this system the ground station sent out two low-frequency signals which overlapped by three degrees. One was modulted with the Morse code letter A (·-) and the other with the code letter N (-·). When he was flying in the overlap area, the pilot would hear both letters with equal loudness and they became a solid tone, forming the radio beam, as it was commonly called. The only airborne equipment needed was a low-frequency receiver and a well-trained ear.

Another early development in radio navigation was the radio direction finder, the RDF. As we mentioned in the section on loop antennas, a loop is highly directional in its reception, and a three or four turn loop was installed across the fuselage, behind the cabin. When the airplane was pointed either directly toward or directly away from the station, the signal was the weakest - the null. If this weak signal got even weaker, the airplane was going away from the station, but if it got stronger, it meant the airplane was flying toward it.

The number of airplanes in our national airspace has steadily grown and with this growth, there has been a greatly intensifed need for a means of navigation that is completely independent of visibility of the ground. Some of the basic requirements for an electronic navigation system are these:

1. The system must be highly accurate, and its inherent accuracy must be in the ground facilities. By this it is meant that not only airliners and executive aircraft should be able to have the ultimate in accuracy, with the use of their elaborate and expensive receiving equipment, but that, at the same time, small private aircraft can use the system with much lower-cost equipment, and still have an adequate degree of accuracy.

2. The system must be relatively simple to use without requiring the exclusive attention of the pilot, allowing him to still have complete control of the instrument flight conditions of the airplane.

3. It must allow maximum utilization of the available airspace.

A. VERY HIGH FREQUENCY OMNIRANGE NAVIGATION SYSTEM [VOR]

Soon after World War II, a system of radio navigation was developed that promised to answer the needs of the burgeoning air transportation in the United States. In this system, a ground station transmits two signals in the very high frequency (VHF) range. One of these signals is amplitude-modulated at 30 hertz and the other is frequency-modulated with a 10 kilohertz subcarrier and is fed into an antenna complex by a device known as a goniometer. This rotates at 1800 RPM and in so doing, produces a signal which appears as 30 hertz AC when received.

The rotating signal and the fixed reference signal are in phase as the goniometer rotates past magnetic north, but as it rotates, the signals get progressively out of phase until, at magnetic south they are 180° apart. Fig. 86 shows the phase relationship of the two signals at four different directions from the station. Anywhere along a line magnetic north of the station, the two signals will be in phase, Fig. 86-A. Along a line magnetic east of the station the signals will be 90° out of phase, Fig. 86-B. Magnetic south of the station, they will be 180° apart, Fig. 86-C, and, at magnetic west, their phase difference will be 270°. They will be back in phase again at magnetic north.

Both of these signals are received on a horizontally-polarized V-dipole antenna and taken into the superheterodyne receiver.

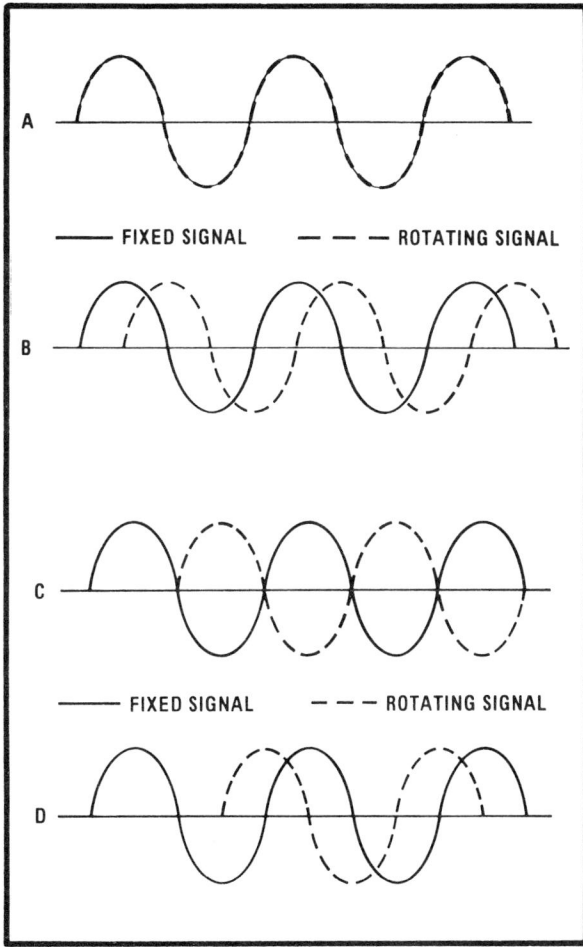

Fig. 86 *VOR Phase Relationships*

A - *Along a line magnetic north of the VOR station, the fixed and reference signals are in phase.*

B - *Along a line magnetic east of the station the signals are 90° out of phase.*

C - *Magnetic south of the station, they are 180° apart.*

D - *Magnetic west of the station, they are 270° out of phase.*

After going through all of the required stages of mixing, converting into an intermediate frequency, amplifying, detecting, and demodulating, the audio portion of the signal is fed into the VOR circuitry. A filter allows the amplitude-modulated 30 hertz reference signal to enter one portion of the circuit, and a 10 kilohertz band-pass filter allows the frequency-modulated rotating signal to enter another part of the circuit. This FM signal is passed through a limiter and a discriminator and becomes 30 hertz AC.

The AM 30-hertz signal is fed into a calibrated

phase shifter which is controlled by the pilot and known to him as an omni bearing selector (OBS). The dial of this phase shifter is calibrated in 360° and looks like the dial of a magnetic compass. If the airplane should be located magnetic north of the station and the phase shifter set on zero (or 360), the two signals would be in phase. And if the airplane should be 217° clockwise from magnetic north, from the station, the two signals would not be in phase until the phase shifter (OBS) was rotated to read 217. After the signal leaves the phase shifter, it is amplified so that it will have the same amplitude as the 30-hertz signal from the discriminator and both of the 30-hertz signals are fed into a phase detector which drives the needle of the course detector, or L-R indicator.

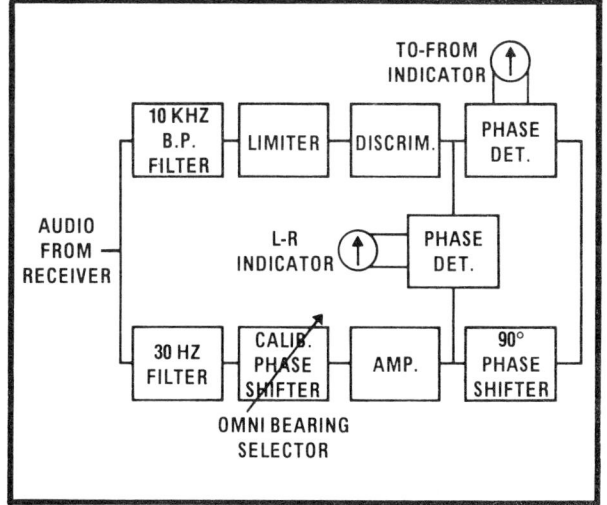

Fig. 87 *Block diagram of VOR receiver.*

There is another phase shifter and phase detector in parallel with the one that drives the course detector, and the reference signal which has already been shifted in order for it to agree with the rotating signal is again shifted; this time by exactly 90°, and again compared with the rotating signal. This shift tells whether the reference signal is leading or lagging the rotating signal, and on the face of the instrument it actuates the *To-From* indicator.

Now for a real quick rundown on the way the VOR is used. When the pilot tunes in the VOR station, the signal is fed into the VOR circuitry, and the course deviation indicator will, in all probability move off to one side or the other. By rotating the omni bearing selector, the pilot can bring the needle to the center. When it centers - and let's assume it does so, when the OBS is reading 185° -

the *To-From* indicator will read either *To* or *From*; and, for our example, let's assume that it reads *From*. This means that the airplane is on a line leading out of the VOR station 185° clockwise from magnetic north - or, as a pilot would say, on the 185° radial, position *A*, Fig. 88. Furthermore it means that if the pilot will turn his airplane to a heading of 185° magnetic, he will be going away *from* the station. If the *To-From* indicator had said *To*, it would have meant that the airplane was on the 005°-radial (the opposite side of the compass from 185°); in that case, if the pilot were to turn the airplane to a magnetic heading of 185°, he would go *To* the station, position *B*, Fig. 88.

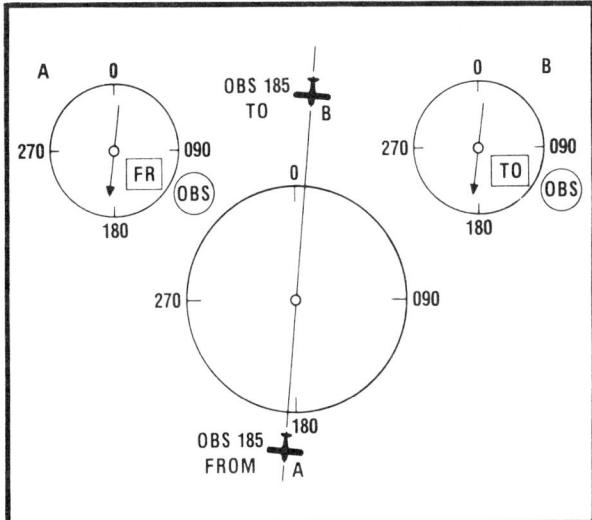

Fig. 88 *When the pilot has the indication shown in A, he will be south of the station, at position A. If his indication is that shown in B, he will be in position B.*

B. AUTOMATIC DIRECTION FINDER [ADF]

VOR operates in the very high frequency range and, as a result, its reception is good only in a line of sight. That is, its reception is unusable at low altitudes when terrain such as mountains are between the airplane and the transmitter. For these conditions, when the VOR is unusable, there is a very handy backup in most airplanes in the form of ADF, or automatic direction finding equipment.

We already know that a loop antenna is highly directional in its reception characteristics, and we have found that it is possible for a sharp pilot to tell by the build or fade of the signal whether he was going toward or away from the transmitter. To make this information easier to use, the

indication of the ADF has been put on a dial, and the 180° ambiguity has been removed from it.

The loop antenna receives the signal with a field pattern such as we see in Fig. 89-A. When the current is maximum in the loop, the induced voltage in its two sides have the opposite polarity. A second antenna is used with the ADF system. This one, an omnidirectional, sense antenna, has a polarity like that seen in Fig. 89-B. When the signals from these two fields are fed into the ADF circuitry they produce a resultant field shaped like a heart, a cardoid pattern.

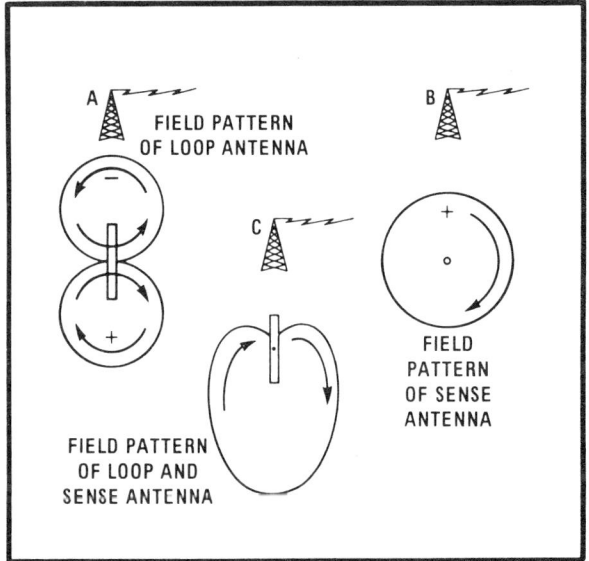

Fig. 89 *Signal pattern of an ADF antenna system*
A - *The loop receives a signal with a field pattern similar to a figure eight.*
B - *The sense antenna receives a signal of equal strength from all directions.*
C - *When the two signals are mixed, the signal from the sense antenna cancels that in one side of the loop, and adds to that in the opposite side, producing a heart shaped, or cardiod, field pattern.*

An ADF receiver normally has a function switch to select the antenna in use. If it is desired to receive weather information, the receiver is placed in the "ANT" position and only the sense antenna is used, so the signal will be received from any direction with equal strength. If the switch is placed in the "LOOP" position, only the loop is in use, and the equipment may be used for manually locating a transmitter. One hundred and eighty-degree ambiguity is present in this mode of operation. In the ADF position, both the loop and sense antenna feed the signal into the circuit.

Fig. 90 is a block diagram of an ADF system.

The signal is received by both the loop and sense antennas, and that from the loop is amplified and combined with the signal from the sense antenna.

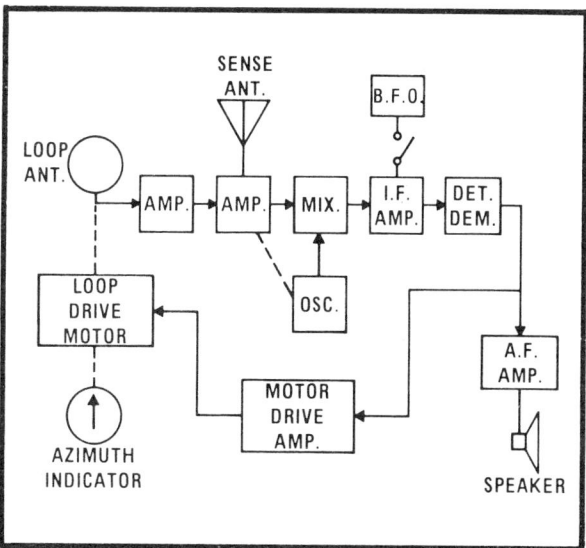

Fig. 90 Block diagram of an ADF receiver.

The signals from the two sources are mixed with a voltage from the local oscillator to produce an intermediate frequency whose strength varies with the position of the loop antenna. This *IF* voltage is further amplified, detected, and demodulated, and the resulting audio frequency is amplified and fed into the speaker. A voltage is taken off just ahead of the AF amplifier and amplified enough to drive the loop motor, which rotates the antenna to the null, or the point of lowest signal strength. The azimuth indicator is electrically driven from the loop drive motor and indicates to the pilot the relative bearing between the null position of the loop and the nose of the airplane.

To use the ADF, the pilot selects the station he wishes to use and identifies it by its call signal. The loop drive circuit automatically rotates the loop until its signal is minimum, and the azimuth indicator tells the pilot the direction of the station from the nose of the airplane. If he wishes to fly to the station, he simply turns the airplane until the indicator points to zero, which tells him that the station is directly ahead of the nose of his airplane; and if he keeps the indicator reading zero, he will eventually arrive over the station.

ADF receivers usually have a frequency range of from about 200 kilohertz up to around 1600 kilohertz. This includes all of the low-frequency nondirectional beacons and the compass locators, as well as the entire commercial broadcast band.

The early ADF loop antennas were large, open loops that were susceptible to the formation of ice and offered considerable wind resistance. As aircraft became faster, the need for streamlining became greater and the loops were made smaller and housed in streamlined plastic "footballs." The air core loops were later replaced with those having powdered iron cores which were much more compact and could be mounted in smaller housings that further cut down the wind resistance.

The most recent development in ADF loops is the fixed loop, almost universally used in the current production ADF systems. Two fixed coils, 90° to each other, are installed in either a flush or streamlined housing. The signal from the transmitter is received by these two coils and is carried into the receiver where two more coils produce identical fields. The fields from these coils produce a resultant field in a goniometer, or resolver, inside the receiver that produces the directional signal.

The fixed loop and the rotating loop antennas are identical in operation; their only difference is in the mechanical details. The rotating element used with the fixed loop is inside the receiver case and can be made of much lighter material, not being subject to rough handling as is the exposed antenna.

The sense antenna used with ADF systems has, throughout the years, been a long wire, usually extending from the top of the cabin to the vertical fin. This exposed wire was one of the weaker elements of the system, being a source of precipitation static, and is continually in danger of being carried away with a load of ice. The most recent developments in ADF systems incorporates the sense antenna in the same unit as the fixed loop. Using this smaller antenna requires a great deal of amplification to compensate for its lower efficiency, but obtaining this amplification is now well within the realm of the modern state-of-the-art electronics.

55. What equipment is necessary to receive the low-frequency radio range?

56. What was the one major problem with the radio direction finder?

57. What type of antenna is used with the VOR?

58. What type of antenna is used with the ADF?

59. What prevents the ADF system having 180° ambiguity?

C. INSTRUMENT LANDING SYSTEM [ILS]

It is not enough to be able to fly from point A to point B if you are unable to land at the airport when you get there. To facilitate landing in conditions of low visibility, the instrument landing system (ILS) has been devised. This is, as the name implies, a complete system consisting of three basic components. The localizer directs the pilot right down the centerline of the runway, the marker beacons locate his position along the extended centerline to tell him how far he is away from touchdown, and the glide slope is a slanting radio beam that allows him to descend at the proper angle to reach the minimum altitude the proper distance from the end of the runway. Included in the complete ILS package are the approach lights, of course, but since they are not airborne components, we will not discuss them here.

1. Localizer

The VOR/Localizer indicator when used with the VOR shows the pilot whether or not he is on the course line he has selected. Each dot is an indication of 2-1/2 degrees off course. In other words, the airplane would be off course ten degrees, if the needle has swung full scale. This is acceptable for flying along the airways, but for the final approach, information of much closer tolerance is required, and the localizer is used to provide the pilot with steering information on final. The same instrument and receiver are used for the localizer as for the VOR, but the portion of the circuitry used is quite different, and the sensitivity of the indicator is such that the full swing of the needle indicates a deviation off the **runway centerline of only 2-1/2 degrees.**

The ground facilities for the localizer are altogether different from those used by the VOR,

Fig. 91 Each dot represents a deviation of one half degree from the centerline of the localizer.

but both systems operate in the same frequency range and therefore use the same antenna and the same front end of the receiver. Terminal VOR's, that is the low-powered transmitters, usually located right on the airport, transmit on frequencies between 108.2 and 111.8 megahertz and are all in the increments of even tenths. In the same range, operating from 108.1 to 111.9 megahertz in odd tenths, are the ILS localizers. The identifying Morse codes used for the localizers and VOR's have the same last letters, but the codes for the localizers are prefixed with I (··). For example, the identifier for Los Angeles International Airport is LAX (·–·· ·– –··–) and one of the ILS localizers there is identified by ILAX (·· ·–·· ·– –··–).

Rather than being a phase comparison system, like the VOR, the localizer is a voltage comparison system. Located at the end of the runway is a system of antennas that radiate two carriers on the same frequency. One is modulated with a 90-hertz tone, the other with 150-hertz, Fig. 92. The one modulated with 150 hertz transmits a sort of a lima bean-shaped pattern on the right side of the runway, as viewed from the normal approach end, and the one with the 90-hertz modulation produces a similar pattern on the left side of the runway.

Since there is only one carrier frequency, the receiver will pick up the signal and process it through the RF and IF stages, and deliver it to the

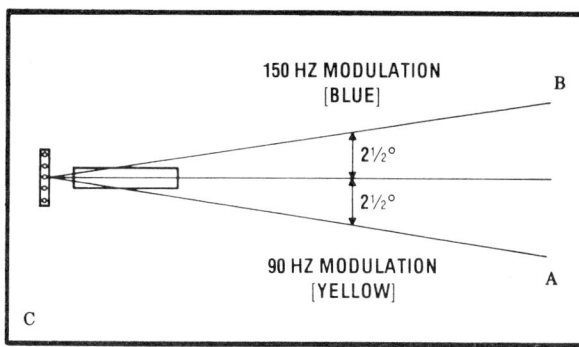

Fig. 92 *The left side of the runway, from the normal approach end, is in the 90 hertz modulated field, while an airplane on the right side will receive 150 hertz modulation.*

AF amplifier. Here two filters, Fig. 93, operate on the signal and the 90-hertz modulation is converted into direct current of a certain polarity. The 150- hertz signal passes through another filter and when rectified becomes DC of the opposite polarity. The two DC voltages are then fed to the course deviation indicator where they are compared. Rather than being calibrated in terms of *Left* and *Right*, the indicator has two colored segments at the end of the indicator needle—the segment on the left side of the dial is blue, and the one on the right is yellow.

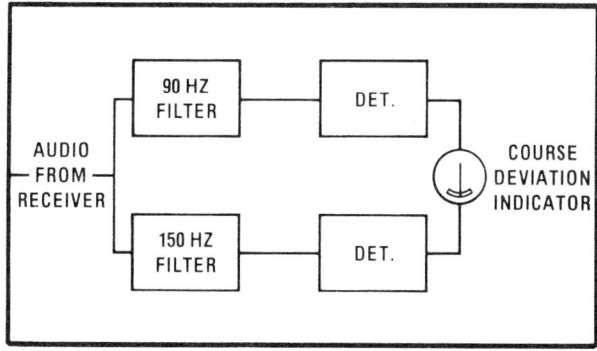

Fig. 93 *The course deviation indicator measures the relative strength of the two modulations to provide localizer information.*

This indicates only the side of the runway on which the airplane is located. Looking back at Fig. 92, we find that if the airplane is in position *A*, it is on the "yellow" side, and will be receiving the 90-hertz modulation stronger than the 150, and the needle will be driven to the right side of the dial. (This does sound backwards, but just bear with me and you will see the logic of the marking).

If the airplane should be on the right side of the runway, at position *B*, for example, the 150-hertz

signal would be the stronger; and when converted into DC it would drive the needle to the left side of the dial, into the blue segment. If the airplane were right down the centerline of the runway, it would be receiving both the 90- and 150-hertz modulation with equal intesity; the signals would cancel each other, and the needle would stay in the center of the dial.

The reason for the blue and yellow segments on the dial and the identification of the sides of the runway as blue or yellow rather than right or left, ties in with the indication of the VOR: In normal use of the VOR, the pilot turns his omni bearing selector until the indicator reads *To*, if he is going toward the station, and *From*, if he is flying away from it. When it is set this way, he will always turn *toward* the needle to get back on the radial if he drifts off.

If the reciprocal heading were selected, he could still use the VOR, but he would have to turn *away* from the needle to get back on the desired course line.

The localizer does not have the capability of correcting reverse-sensing, and it is set up in such a way that when approaching the runway from the normally used end (front course), the pilot will have normal sensing of the needle. This means that if he is in position *A*, Fig. 92, he will be to the left of the runway, but the needle will be off to the right, Fig. 92-B, he will turn *toward* the needle to get back on centerline. Now if he approaches the runway from the opposite direction (back course), he will have reverse-sensing of the needle, and if he finds himself in position *C* Fig. 92, the needle will again be in the yellow. This time, however, since he is on a back course, he knows that he must turn *away* from the needle to get back on the centerline.

Even though the operating principles of the two systems are quite different, they both operate in the same frequency range, and both signals are horizontally polarized; so the antenna is the same for the localizer as is used for the VOR.

2. Compass Locators

For an ILS approach to be made, the pilot must find the localizer. This is normally done by the use of an outer marker locator which is a low-frequency nondirectional radio beacon, located about five miles from the approach end of the runway at the same place as the outer marker. This is tuned in on the ADF receiver and the airplane flown to the locator, at which time the ADF needle will swing from pointing ahead of the airplane to pointing behind it. When this outer marker locator has been reached, the pilot turns away from the airport on the heading given on the approach chart and flies for a specified time. He then makes what is known as a procedure turn, which allows him to intercept the localizer, headed inbound. He can then fly down the localizer centerline to the runway.

There is another compass locator, called the middle marker locator (LMM), about a half-mile from the end of the runway. After the outer marker is passed, flying inbound, the pilot can tune the middle marker frequency on the ADF and fly to it. When the ADF needle swaps ends here, he knows that he is within about a half-mile of the end of the ILS runway.

3. Marker Beacons

Compass locators are used, as the name implies, to locate the markers, but for the actual use on an ILS approach, a more precise method of position determination is needed. This is supplied by marker beacons, located at the same place as the compass locators.

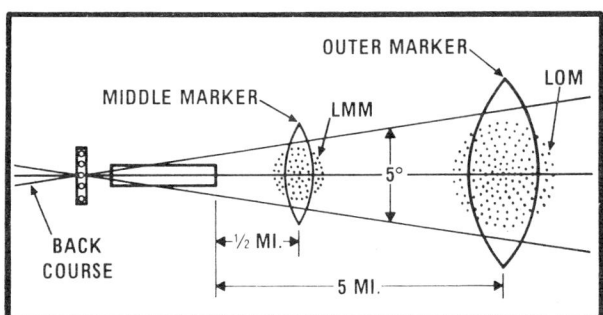

Fig. 94 Compass locators and marker beacons are co-located. The compass locator is received on the ADF, and the marker beacon on a fixed frequency 75 MHz receiver.

The marker beacon uses a single-frequency transmitter, with a carrier of 75 megahertz, and a power of two to three watts. This signal is fed into a highly directional antenna, which transmits its signal straight upward in a fan shaped pattern. The outer marker is modulated with a series of 400-hertz dashes, at a rate of two per second. The middle marker is modulated with a series of 1300 hertz alternating dots and dashes, and the

Airways, or Z marker, is modulated with a series of 3000-hertz dots. These 3000-hertz markers may be placed at the point where the decision height should occur on an ILS approach, or they may be used to indicate the point on a back course where the approach descent should begin. Those markers used on the back course, instead of a series of dots, are modulated with a series of two-dot combinations.

The receiver for the marker beacon is a fixed frequency superheterodyne, Fig. 95.

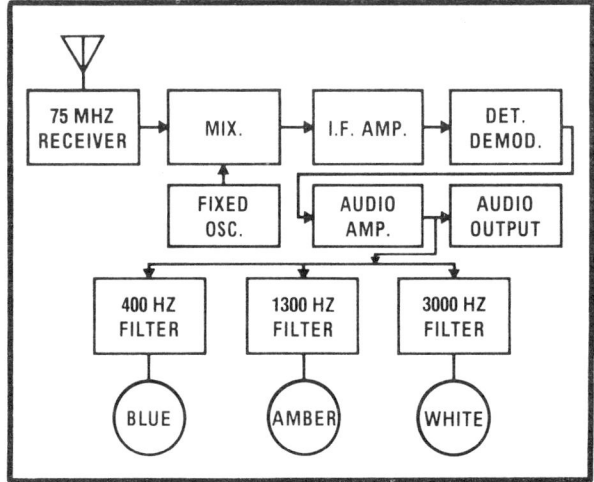

Fig. 95 Block diagram of a marker beacon receiver.

The signal from the marker beacon is received and carried through the RF, IF, and AF stages; but instead of going to the speaker, it is sent to a series of filters. If the signal received is modulated with 400 hertz, it feeds through the 400 hertz filter and causes a blue light to illuminate on the instrument panel, Fig. 96.

Fig. 96 Marker beacon lights.

If the airplane is over the middle marker, the 1300 hertz signal is filtered out and turns on the amber light. Any time the 3000-hertz signal is received, it is filtered and causes the white light on the panel to light up.

The fixed frequency of the marker beacon receiver allows its antenna to be less complicated than that required for a tunable receiver that must accomodate a wide range of frequencies. There are three types of antennas that may be used with the marker beacon receivers: A wire, a rod, and the newer, completely enclosed, "canoe" antenna, Fig. 97.

Fig. 97 *Seventy five megahertz fixed frequency marker beacon antenna.*

4. *Glide Slope*

The compass locator is used to find the localizer, the localizer directs the pilot down the runway centerline, the marker beacons identify the pilot's position along the localizer, and the glide slope establishes the correct descent for the approach.

Like the localizer, the glide slope transmits two signals on the same frequency, one modulated with 90 hertz, and one with 150 hertz. The carrier for the glide slope is in the UHF band from 329.3 to 335.0 megahertz. Instead of requiring the pilot to tune the glide slope receiver, each ILS has the glide slope paired with its localizer so that when the pilot tunes the localizer frequency, the glide slope receiver is automatically tuned to its proper frequency.

The antenna pattern of the glide slope is positioned in such a way that the 150-hertz modulated signal is transmitted below the 90-hertz signal, and they overlap by about 1-1/2°. This overlap occurs at a nominal 3° above the horizontal, Fig. 98.

As the pilot makes his approach, he flies at the altitude specified for the approach, and while he is outside of the outer marker, the glide slope indicator needle will point upward, indicating that the glide slope is above him. At the outer marker, the pilot should intercept the glide slope and the needle should come to the center, Fig. 98. The pilot then begins his descent and keeps the needle

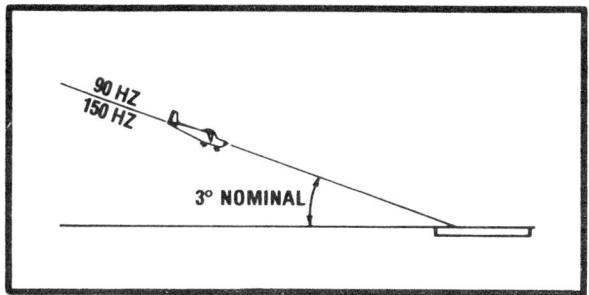

Fig. 98 *The glide slope projects upward from the surface, a nominal three degrees.*

centered. If he should go above the prescribed glide slope, the needle will **point downward**; below the slope, the needle will **point up**.

Fig. 99 *The horizontal needle on the VOR/Loc indicator is the indicator for the glide slope.*

The circuitry for the glide slope is essentially similar to that used for the localizer. A twenty-channel crystal-controlled receiver is tuned with the same control that selects the localizer frequency. Ninety and 150-hertz filters select the modulation being received and convert into direct current. The DC from the two filters is compared, and the output drives the needle either up or down.

The null, or point of equal signal, occurs not only on the correct gilde slope, but because of the reflection of the signal from the earth, will produce several nulls, or on-course signals, above the true slope. There should be no danger from these false glide slopes, because they are all above, and steeper, than the true slope. On a properly executed ILS approach, the first time the glide slope centers is at the time the airplane passes the outer marker, at the altitude specified on the approach plate.

The antenna for the glide slope is a UHF dipole, and since the length of the antenna varies inversely as the frequency, this antenna is much shorter than the VHF dipole used for the localizer and the VOR. Some glide slope antennas are mounted inside the cabin, some above or below the fuselage, or in the nose, and some are mounted in the front of the VOR antenna in the form of a "Cat's whisker".

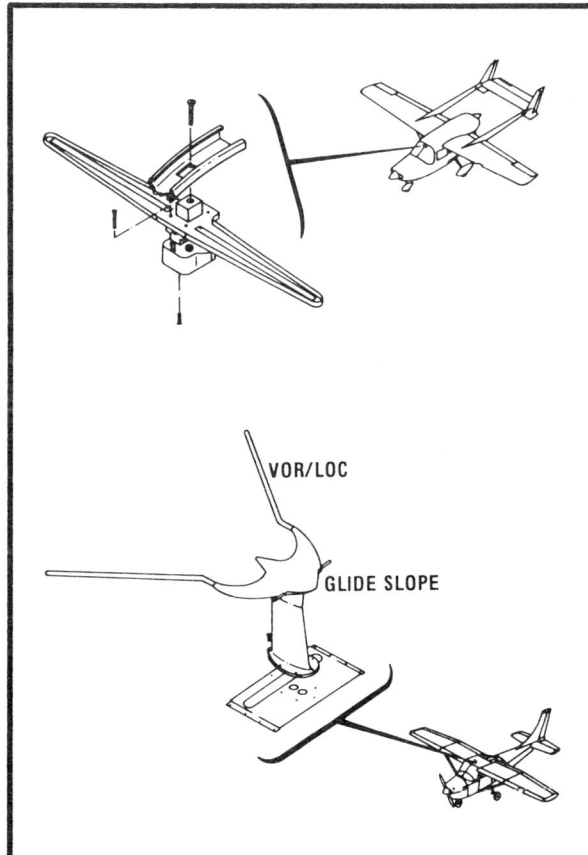

VOR/LOC

GLIDE SLOPE

Fig. 100 Glide slope antenna.

QUESTIONS

60. Name the three airborne portions of an instrument landing system.

61. How does the sensitivity of the localizer compare with that of the VOR?

62. What receiver is used to receive the signal from the compass locator?

63. On what frequency do marker beacon transmitters transmit?

64. What type of antenna is used to receive the signal from the glide slope transmitter?

65. How is the glide slope receiver tuned?

D. DISTANCE MEASURING EQUIPMENT
[DME]

The problem of getting directional information from a ground station has been solved for general aviation aircraft by the use of VOR. But in order to get the *distance* from the station, a portion of a complex military navigational system is used. TACAN, or *Tactical Air Navigation*, is a pulse system that requires very sophisticated reception equipment to use the direction-and distance-information furnished by the ground station, but civilian aircraft have been allowed to use the distance measuring portion and have in it a useful instrument call DME, or Distance Measuring Equipment.

The principle of DME is quite simple, but only by the use of very complex electronic circuitry can the desired result be achieved. Fig. 101 is a block diagram of the airborne and ground equipment.

A signal, consisting of about 150 random-spaced pairs of pulses, is generated and used to modulate the transmitter. These pulses are also fed into the search-and-track portion of the set and stored in a memory system until needed. The transmitter puts the modulation on a carrier of between 962 and 1024 megahertz, or between 1151 and 1213 megahertz.

The modulated carrier is fed to the preselector which acts as a switch connecting the antenna alternately to the transmitter or to the receiver. When the preselector is connected to the transmitter, the modulated signal is transmitted from the airplane. The ground station picks up the signal, and its preselector directs it into the receiver, then into a video, or wide band amplifier, and into a delay circuit where it is held for a specific time period and then released into the modulator. Here it is amplified and modulates the transmitter which uses a carrier of between 1025 and 1150 megahertz to return it to the airplane when the preselector connects the

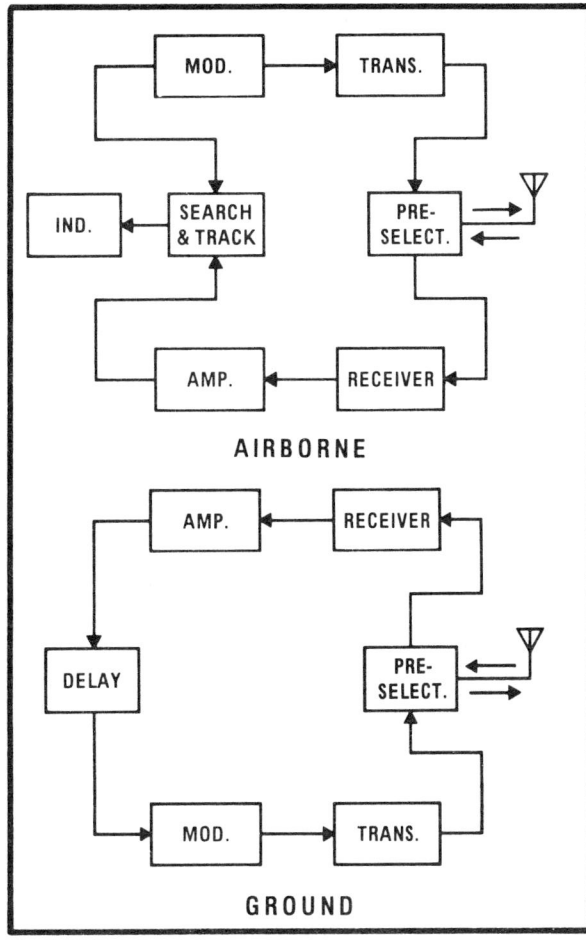

Fig. 101 Block diagram of distance measuring equipment.

transmitter to the antenna.

The airborne antenna picks up the signal, and the preselector directs it into the receiver and the amplifier, and then into the search and track circuits. The received signal is compared with the original that has been stored by the modulator, and if they do not coincide, the search circuit will continue to hunt for the proper pattern of pulses. The indicator, during the search, will move out to the end of its range and then back toward zero, and repeat this until it finds the sequence it transmitted. When this particular pattern is found, the search-and-track circuit locks on and tracks it. The number of pulse pairs sent out drop from about 150 per second to around 30 per second, and the indicator shows the distance in nautical miles from the station.

The correlation between time and nautical miles is based on the assumption that a pulse of electrical energy travels through the air at a rate of close to

161,900 nautical miles per second. The time required for the pulse to leave the airplane, reach the ground station, delay for a specific period of time, and return to the airplane can be read directly in nautical miles to the station. This is naturally a slant-line distance and does not take into consideration the altitude of the airplane above that station.

DME is a pulse system, meaning that it transmits and receives pulses of UHF electrical energy, and it requires a short, vertically polarized antenna - either a blade or a stub such as that shown in Fig. 102.

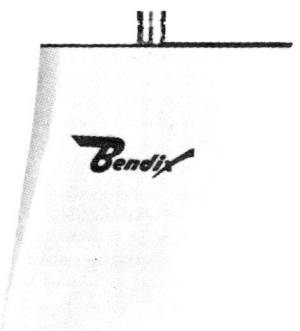

Fig. 102 A vertically polarized UHF antenna is used for both the distance measuring equipment and the radar beacon transponder.

E. RADAR BEACON TRANSPONDER

The one piece of electronic equipment in the cockpit that appears to do the least, yet serves an extremely useful role in safe flying, is the radar beacon transponder.

The primary radar on the ground sends out pulses of extremely high-frequency, high-power energy that radiates outward in space and does not bounce back from the ionized layer of the atmosphere that surrounds the earth. But when they strike a metal object such as an airplane, they do bounce back. The energy that returns to the radar, even though in the nature of microwatts, is enough, after the proper amplification in the receiver, to leave a trace on the phosphorescent radar scope being watched by the traffic controller.

The controller must sort out the targets on his scope and identify them by having the pilots make the appropriate radar-identifying turns; once the plane is identified, a small plastic "Shrimp boat" is placed over the dot as it progresses across the scope. Until recently there has been no way of determining the altitude of the airplane being followed, or of positively and instantly identifying it to distinguish it from other airplanes that are close together on the scope.

Elaboration on a system devised in World War II, the Identification, Friend or Foe (IFF), is being used successfully today to overcome the limitations of the primary radar with regards to positive identification of the airplanes being followed.

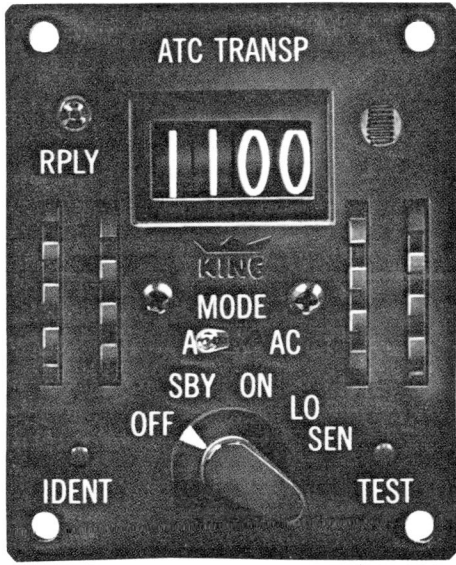

Fig. 103 Radar beacon transponder.

A special interrogator antenna is attached to the primary radar antenna and rotates with it. This interrogator sends out a pulse of energy on a frequency of 1030 megahertz in a highly directional pattern. The aircraft receives this interrogation, and transmits a coded response on a frequency of 1090 megahertz. This response is received on the ground by the primary radar, and instead of the airplane appearing on the scope as the the small return, it appears as a bright single or double slash because of the energy transmitted from the airplane.

The pilot selects the code requested by the controller, and all of the airplanes flying in a particular level of the airspace, or operating on a certain type of clearance, may be quickly sorted out by the controller. Much safer flight can thus be maintained along our crowded airways and terminal areas.

In order to positively identify an airplane, the controller may ask the pilot to "squawk ident". The pilot responds by momentarily pressing the *Ident* button on the control head, and for a period of about twenty seconds, a special identification pulse is sent out each time the transponder is interrogated. An ident signal causes the return on the controller's scope to be filled in between the two slashes and immediately and positively identifies the airplane.

If a pilot encounters an emergency situation and wants to notify the ground station of his problem, he selects the code 7700. When the ground station receives this reply to its interrogation, both slashes become wider and brighter, and the controller can take any necessary action to aid the pilot in his distress.

The transponder is capable of replying to the ground interrogation in any of 4,096 codes.

The operation of the transponder is similar to that of the DME previously described. Follow Fig. 104 and see that the interrogation signal on 1030 megahertz is received by the antenna, passed through the preselector, then amplified and detected by the video amplifier. It then passes into the decoder which is set to respond only to the interrogation pulse selected by the pilot's control. The output of the decoder is fed into a keyer which produce a pulse, and this in turn causes the encoder to produce a series of pulses appropriate to the code selected.

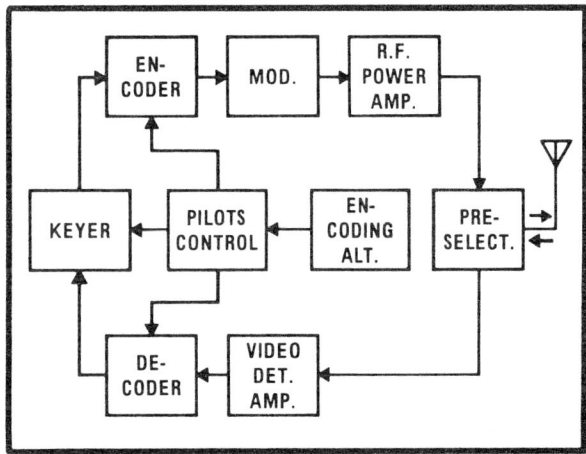

Fig. 104 Block diagram of a radar beacon transponder.

These pulses modulate the 1090 megahertz carrier, which is amplified in the RF amplifier.

The preselector then directs this energy into the antenna and prevents any of this signal from entering the receiver portion of the circuit.

When the transponder is operating in mode A, it provides identification information only; but when the pilot selects Mode C, the interrogation is answered by a code produced in an encoding altimeter and responds with information that produces a read-out on the controller's radar scope telling the altitude of the airplane in one-hundred-foot increments.

Many of the modern encoding altimeters are of the optical type in which the altimeter mechanism drives a disc having some segments transparent and some opaque. A light shines through the disc and energizes a photoelectric cell. For each one-hundred-foot increment a different code is sent to the decoder, and a pulse is formed that will reply to the ground interrogation with the requested altitude; that is, altitude referenced from *standard* sea level pressure of 29.92 inches of mercury.

The only indication that the pilot has of his transponder operating is the winking light on the face of the control head. This light blinks each time the transponder responds to an interrogation from the ground radar.

Installation of the transponder is similar to that of DME. Some small transponders fit into the instrument panel, and others have only the control on the panel and the actual unit itself remotely located. The antenna is a short blade or stub antenna and is located on the belly of the airplane as far as practicable from any other antenna, and in a position that will not be shielded by the landing gear when it is extended.

QUESTIONS

66. What type of antenna is used for the DME?

67. What is indicated by the flashing light on the control head of the radar beacon transponder?

68. What does the pilot do to positively identify himself when using the radar beacon transponder?

69. What is used to code the transponder to provide altitude information to the controller?

70. What type of antenna is used for the radar beacon transponder?

SECTION: VII

Electronics Installation

A. ELECTRICAL CONSIDERATIONS

1. Load Analysis

Before any electronic equipment is installed, it is imperative that the aircraft electrical system be of sufficient capacity to carry the load, and that all of the wiring be of sufficient size that it will not be overheated by the current, or produce an excessive voltage drop.

If the total connected electrical load can exceed the rated output of the alternators or generators, it must be either decreased to within the rated load, or provisions made for the pilot or flight crew to easily manage it and hold it within the rating of the power source. When a storage battery is part of the load, the voltage must be high enough for it to remain continuously charged—except, of course, for intermittent drains placed on the system by the starter, the landing gear motor, or other heavy-current requirements.

If the load is capable of exceeding the source output, there are three acceptable methods of monitoring it:

1. Suitable placards may be put in plain view of the pilot so he can control the equipment and operate only that combination which is within the capability of his electrical system.

2. If an ammeter is installed in the battery lead, and a system voltmeter is installed which reads the voltage on the main bus, as long as the ammeter indicates a charge, and the voltmeter indicates normal system voltage, the source is not overloaded.

3. If an ammeter is installed in the generator or alternator lead, and it never exceeds 100% of the source rated current, the alternator or generator is not overloaded.

If the load requires alternating current and is connected to an inverter, be sure that is does not exceed the rated output of the inverter, and that the load circuit protective device, a circuit breaker or fuse, will open the circuit before the inverter protector trips.

2. Circuit Protection

The wiring must be protected by a fuse or a circuit breaker placed as close to the source as is practicable. The manufacturer of the equipment will normally specify the size of the wire used to supply the equipment and the rating of the circuit protector.

The circuit protector *must* open the circuit before the wire begins to smoke.

Fig. 105 is a list of the approved circuit breaker and fuse sizes that will protect copper aircraft wire. If the manufacturer has not specified which to use, normally either will do; but it must be remembered, that a fuse will usually open the circuit more quickly than a circuit breaker of the same rating.

Figures in parentheses may be substituted where protectors of the indicated rating are not available.

Wire AN gauge copper	Circuit breaker amp.	Fuse amp.
22	5	5
20	7.5	5
18	10	10
16	15	10
14	20	15
12	25 (30)	20
10	35 (40)	30
8	50	50
6	80	70
4	100	70
2	125	100
1		150
0		150

Fig. 105   Wire and circuit protector chart.

3. Wiring

Unless otherwise specified by the manufacturer, electronic installations are normally made using MIL-W-5086 copper wire or, in some instances where the current is high and the distance is long, with the proper size of MIL-W-7072 aluminum wire. Both of these wires are insulated with polyvinylchloride having a nominal breakdown rating of 600 volts.

When selecting the size wire for an installation, you must be sure that the wire size is sufficient to carry the current without overheating, and that the voltage drop will not be excessive. Fig. 106 gives the current-carrying capability of both copper and aluminum wire. When substituting aluminum for copper, you will find that you must use a wire of two wire numbers larger for the same current. For example, a size 8 copper wire will carry about as much current as size 6 aluminum. Because of its more difficult handling character-

istics, it is not usually recommended that aluminum wire smaller than a size 6 be used.

In order to determine the length of wire for the allowable voltage drop, use the wire chart of Fig. 107. If the installation is in a 200-volt system, the voltage drop allowable for the wiring is 7 volts continuous, or 14 volts intermittent. For a 115-volt system, the drop allowed is 4 continuous and 8 intermittent. For 28 volts, a 1-volt-continuous or 2-volt intermittent drop is allowed, and for a 14-volt system, you are allowed only 0.5 volt-continuous and 1-volt intermittent. This allowance is shown by the four scales on the left side of the chart.

Let's take an example of the use of this chart: We wish to install a piece of equipment that requires a continuous current of 30 amperes. The length of the wire we will need is 50 feet, the installation is in a 28-volt airplane, and the wire is to be installed in a bundle. First locate the diagonal line representing 30 amperes and the horizontal line

COPPER Wire size— Specification MIL-W-5068	Single wire in free air—maximum amperes	Wire in conduit or bundled—maximum amperes	Maximum resistance—ohms/1,000 feet (20°C.)	Nominal conductor area—circular mills	Finished wire weight—pounds per 1,000 feet
AN-20	11	7.5	10.25	1,119	5.6
AN-18	16	10	6.44	1,779	8.4
AN-16	22	13	4.76	2,409	10.8
AN-14	32	17	2.99	3,830	17.1
AN-12	41	23	1.88	6,088	25.0
AN-10	55	33	1.10	10,443	42.7
AN-8	73	46	.70	16,864	69.2
AN-6	101	60	.436	26,813	102.7
AN-4	135	80	.274	42,613	162.5
AN-2	181	100	.179	66,832	247.6
AN-1	211	125	.146	81,807	— — — — —
AN-0	245	150	.114	104,118	382
AN-00	283	175	.090	133,665	482
AN-000	328	200	.072	167,332	620
AN-0000	380	225	.057	211,954	770

ALUMINUM Wire size— Specification MIL-W-7072	Single wire in free air—maximum amperes	Wire in conduit or bundled—maximum amperes	Maximum resistance—ohms/1,000 feet (20°C.)	Nominal conductor area—circular mills	Finished wire weight—pounds per 1,000 feet
AL-6	83	50	0.641	28,280	— — — — —
AL-4	108	66	.427	42,420	— — — — —
AL-2	152	90	.268	67,872	— — — — —
AL-0	202	123	.169	107,464	166
AL-00	235	145	.133	138,168	204
AL-000	266	162	.109	168,872	250
AL-0000	303	190	.085	214,928	303

NOTE: *Aluminum wire.* It will be noted that the conductor resistance of aluminum wire and that of copper wire two numbers higher are similar. The use of aluminum wire sizes smaller than No. 6 is not recommended.

Fig. 106 Copper and aluminum wire current carrying capacities.

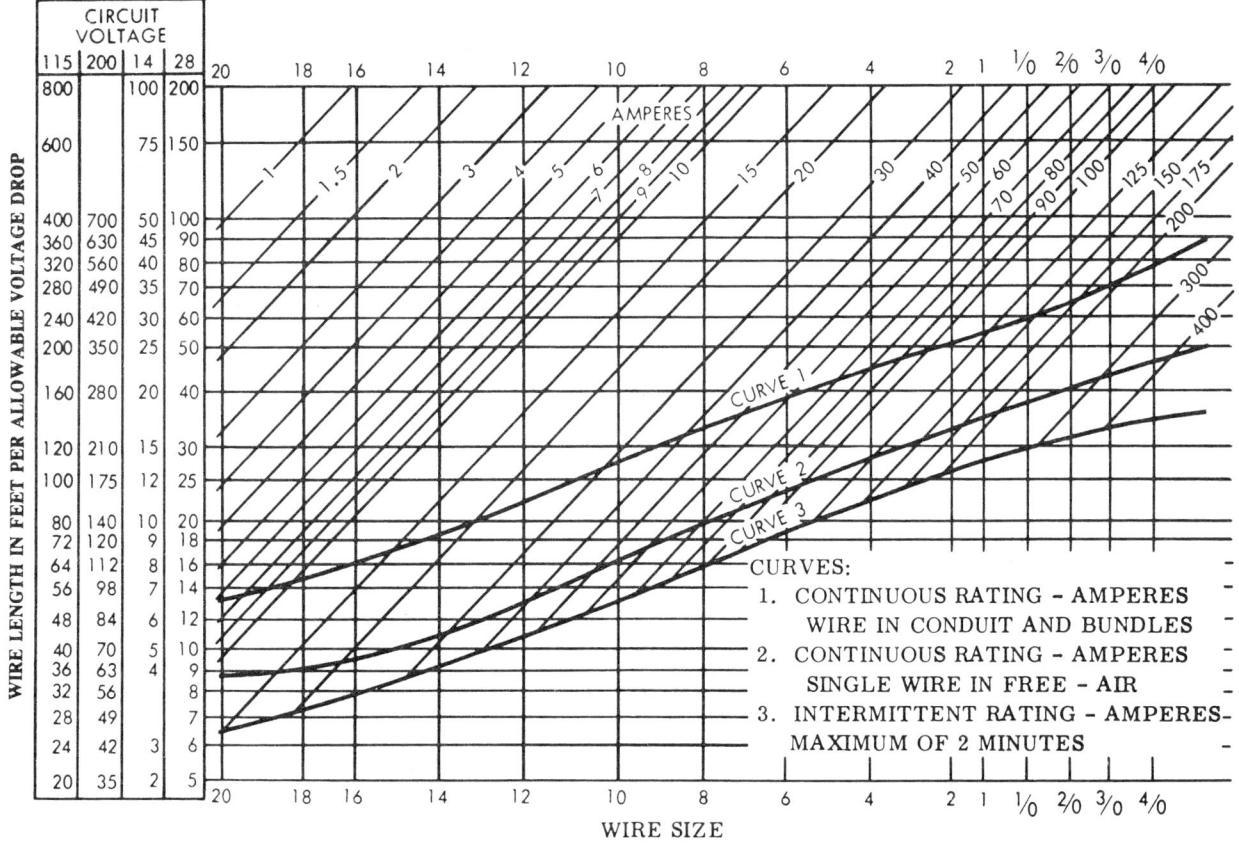

CIRCUIT VOLTAGE			
115	200	14	28
800		100	200
600		75	150
400	700	50	100
360	630	45	90
320	560	40	80
280	490	35	70
240	420	30	60
200	350	25	50
160	280	20	40
120	210	15	30
100	175	12	25
80	140	10	20
72	120	9	18
64	112	8	16
56	98	7	14
48	84	6	12
40	70	5	10
36	63	4	9
32	56		8
28	49		7
24	42	3	6
20	35	2	5

CURVES:
1. CONTINUOUS RATING – AMPERES
   WIRE IN CONDUIT AND BUNDLES
2. CONTINUOUS RATING – AMPERES
   SINGLE WIRE IN FREE – AIR
3. INTERMITTENT RATING – AMPERES-
   MAXIMUM OF 2 MINUTES

WIRE LENGTH IN FEET PER ALLOWABLE VOLTAGE DROP

AMPERES

WIRE SIZE

Fig. 107 Electric Wire Chart.

representing 50 feet in the 28-volt column. These two lines intersect between the vertical lines representing 8 gage and 6 gage, and since we always want to be safe , we will choose the 6 gage. Since this intersection is above curve 1, the wire is of sufficient size that the current will not be excessive for the wire to be installed in a bundle. To verify our choice, we can refer to the chart in Fig. 106, and here we find that MIL-W-5086, AN-6 wire is approved for up to 60 amperes in a bundle, and the required 50 feet of this wire will weigh just a little over five pounds.

4. Wire Termination

a. Terminal Strips

If the wires are to be terminated on a terminal strip, a swaged-on, pre-insulated terminal is normally used. To install one of these, first strip the wire back the length of the barrel of the terminal. Be sure to use the correct size wire stripper so you will not nick any of the wire strands. Slip the terminal over the end of the wire,

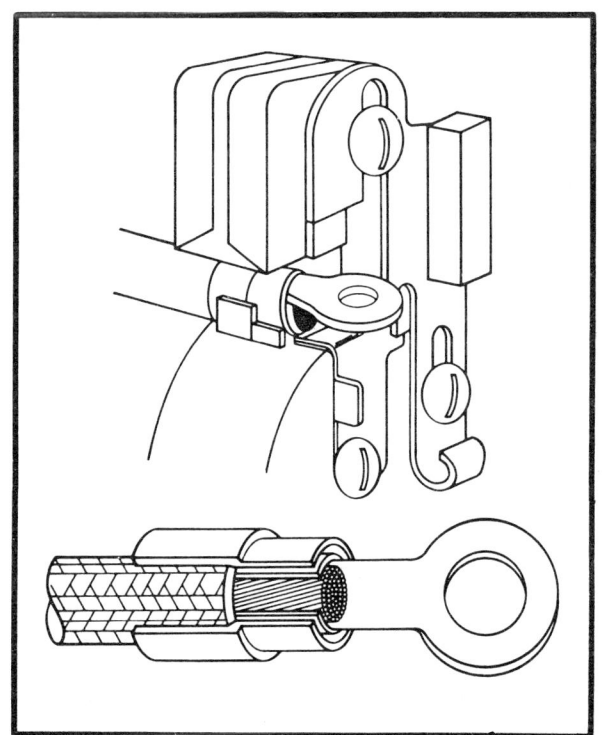

Fig. 108 Proper installation of a solderless wire terminal on aircraft electrical wire.

7-3

making sure the end extends through the barrel, and then crimp the terminal, using the proper tool. Finally crimp the insulation grip over the wire insulation.

The color of the insulated terminal indicated the wire size it will fit. A red terminal fits wire sizes 22 through 18, blue fits 16 and 14, and the yellow terminal fits 12 and 10 gage wire.

When installing wires on a terminal strip, never stack more than four terminals on one stud, and then be sure that they are arranged as those in Fig. 109. If you wish to use a teminal strip as a bus bar with a large number of terminals tied together, use a special connecting bar across the required number of studs and limit the number of terminals on each stud.

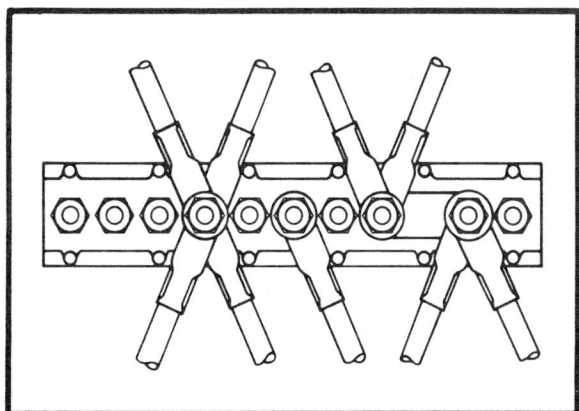

Fig. 109 Proper installation of wires to a terminal strip. Never stack more than four terminals on one stud.

b. Connector Plugs

When wires must be connected and disconnected periodically, connector plugs similar to those shown in Fig. 110 are used.

These plugs may have the wires soldered into the pots, or special pins or sockets may be crimped onto the wire and then pulled into plug. Fig. 111 illustrated the correct way to solder a wire into a plug. The wire is stripped so it will reach to the bottom of the pot and have about 1/32 inch of bare wire sticking out of the top. Slide a short piece of polyvinylchloride tubing over the wire and heat the outside of the pot with the proper size soldering iron. Fill the pot almost full of the correct solder; normally 60/40 resin-core solder is used. Slip the bare end of the wire into molten

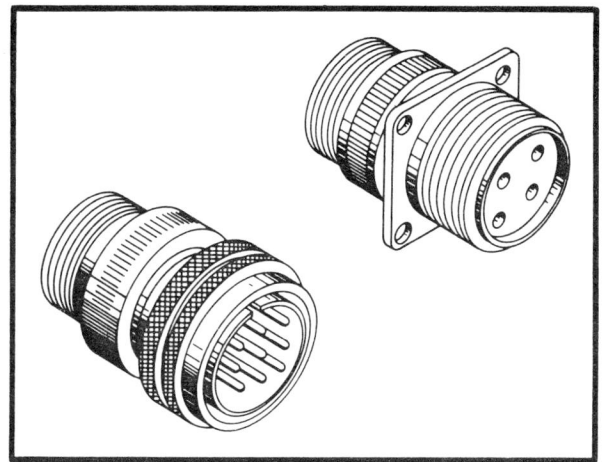

Fig. 110 When it is necessary to connect and disconnect wires periodically receptacles and plugs are used.

solder, and when it is firmly on the bottom, hold it still and remove the iron. As soon as the solder loses its gloss, it will be hard enough for you to release the wire. The solder should flow up around the lip of the pot, but not run down the side, nor should it wick up the wire and make it stiff between the pot and insulation.

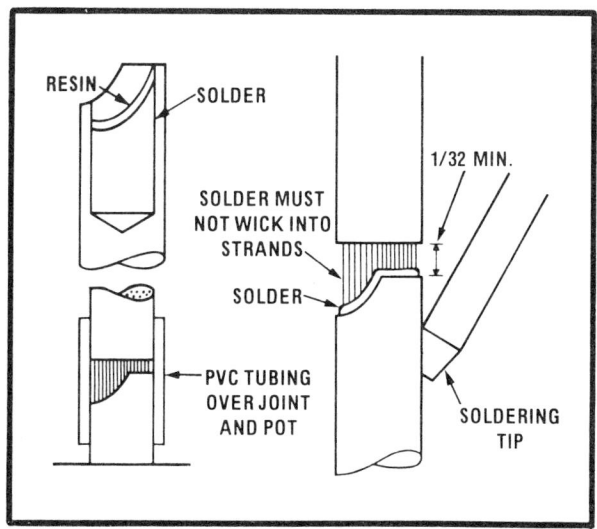

Fig. 111 Proper method of soldering a wire into the pot of a connector plug or receptacle.

A newer generation of connector plugs, designed for faster and more efficient production, uses pins and sockets crimped onto the wires and then pulled or pushed into the plug. These pins and sockets may be of either the front release or the rear release type, Fig. 112.

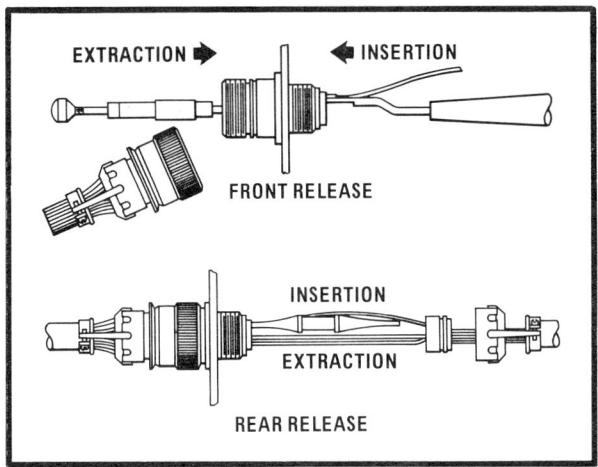

Fig. 112 Plugs and receptacles using pins and sockets crimped onto the wires makes high speed production easier. A special insertion tool is used to assemble the pins or sockets into the body.

QUESTIONS

71. What size fuse would be used to properly protect a 16 gage copper wire?

72. What is the maximum current a 20 gage copper wire can carry in a bundle?

73. What size MIL-W-7072 aluminum wire would properly replace a piece of 2 gage MIL-W-5086 copper wire?

74. What size copper wire would be properly used to carry 20 amperes in a 14 volt system if the wire must be 15 feet long and routed in a bundle?

75. What would be the proper color of pre-insulated terminal to use on a piece of 14 gage MIL-W-5086 wire?

5. Shielding

The input circuits of many pieces of electronic equipment have a high impedance, and as a result almost any stray electrical fields will generate enough voltage to appear as a signal and cause interference with the true signal. For this reason, wires carrying alternating current or those wires bringing signals into the equipment are often shielded. Shielded wire has, in addition to the insulation around the conductor, a braided sheath that completely encloses the wire. This shielding

is grounded to the airframe structure, and any extraneous signal that might get into an input wire or escape from a power lead will be picked up by the shielding and carried to ground. Some circuits are so critical that the current that flows between the shield and the ground will create enough of a field that provisions must be made to prevent it. This condition is called a ground loop, and to preclude it, the shielding is grounded at one end only. Any circuit that requires this extra care will be specified on the wiring installation drawing.

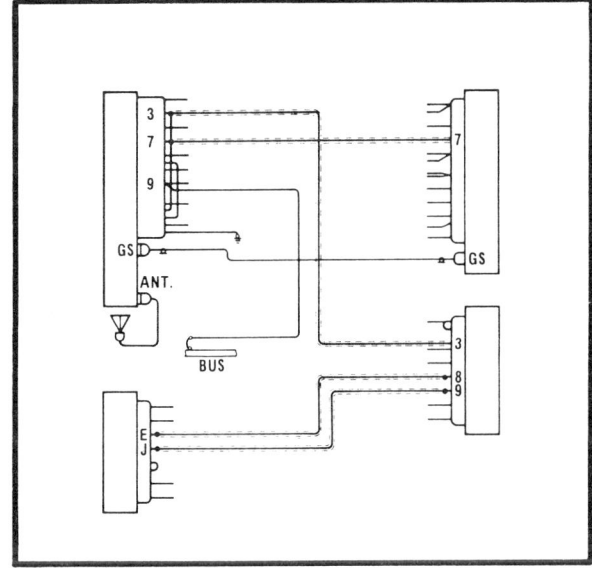

Fig. 113 Some shielded wire has the shielding grounded at one end only to prevent current flowing through the shield and the ground.

6. Bonding

To prevent unwanted potential buildups and to serve as a ground return for shock-mounted equipment, most electronic equipment is bonded. Bonding is essentially the connection of the equipment to ground, normally using a braid or some type of flexible lead.

When installing bonding jumpers, be sure that they are as short as practicable and have the minimum resistance. A resistance of about three milliohms is considered maximum, and if it carries very much ground-return current, care must be taken that there is no appreciable voltage drop in the bonding connection; if a shock-mounted part is bonded, be sure that no strain is placed on the bond when the unit flexes the full amount allowed by the shock mounts.

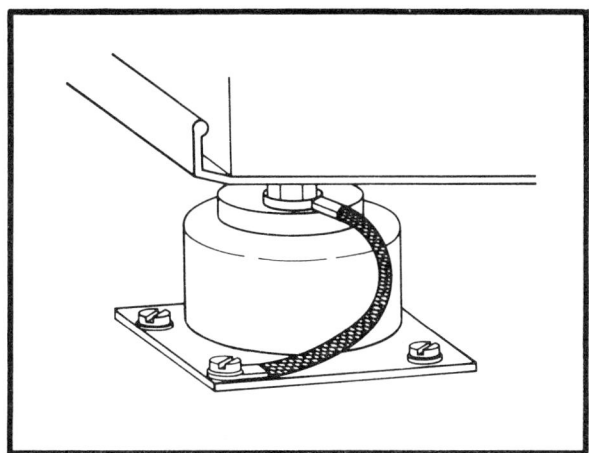

Fig. 114 A bonding braid connecting a shock mounted component to the airframe structure must allow the full movement of the mount.

Since the bonding braid carries current, special care must be taken to prevent its flowing through dissimilar metals, which will cause corrosion. Aluminum alloy jumpers are used for most installations involving aluminum alloy structure and an aluminum alloy component. Cadmium-plated copper is used to bond stainless steel, cadmium-plated steel, or brass. If it is impossible to avoid dissimilar metal junctions, be sure to use a bonding jumper more susceptible to corrosion than the structure to which it is bonded.

Before connecting a jumper to an aluminum alloy part that has been anodized, the oxide coating that protects the metal must be removed. After the connection is made, a protective coating of primer or paint should be applied.

When connecting a bonding braid to a shielded wire, do not use a solder connection, as it will damage the insulation of the wire. Special crimped-on ferrules are available for this purpose.

7. Bundling and Routing

Wires coming from either a terminal strip or a connector plug should all be laid parallel and tied in a bundle. One of three methods commonly used is tying them with a plastic Ty-Rap, a patented strap, wrapped around the bundle and secured with a special tool which tightens the strap and cuts off the loose end.

Wire bundles may also be tied with spot ties, Fig. 115-A, made of waxed linen or waxed nylon. With

this method a clove hitch is tied around the wire and pulled tight, then secured with a square knot. (The advantage of spot ties over lacing is that the individual ties may be cut without loosening the entire bundle.)

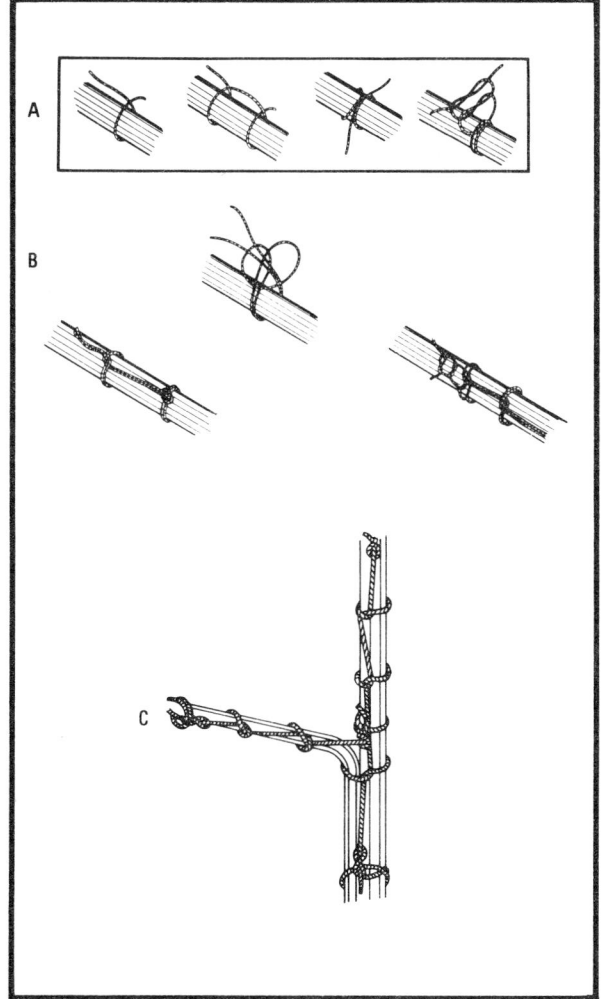

Fig. 115 Wire bundling methods.
A - Individual spot ties.
B - Double string lacing
C - Branching-off lacing.

Lacing is popular because of its speed. A bowline on the bight is used to start the lacing, then half hitches are thrown around the bundle, Fig. 115-B, until the lacing is to be terminated; at which time a clove hitch (two half-hitches) is used, secured with a square knot. If it is desired to branch off of the main bundle, a half hitch is tied just below the branch, and a starting bowline is tied on the main bundle in the fork of the branch. Fig. 115-C.

Wire bundles are secured to the structure with cushioned clamps attached in ways similar to those shown in Fig. 116. Any time the bundle goes through a bulkhead or frame, be sure that it is centered in the hole, and if there is any possibility of chafing, the edges of the hole should be protected with a grommet.

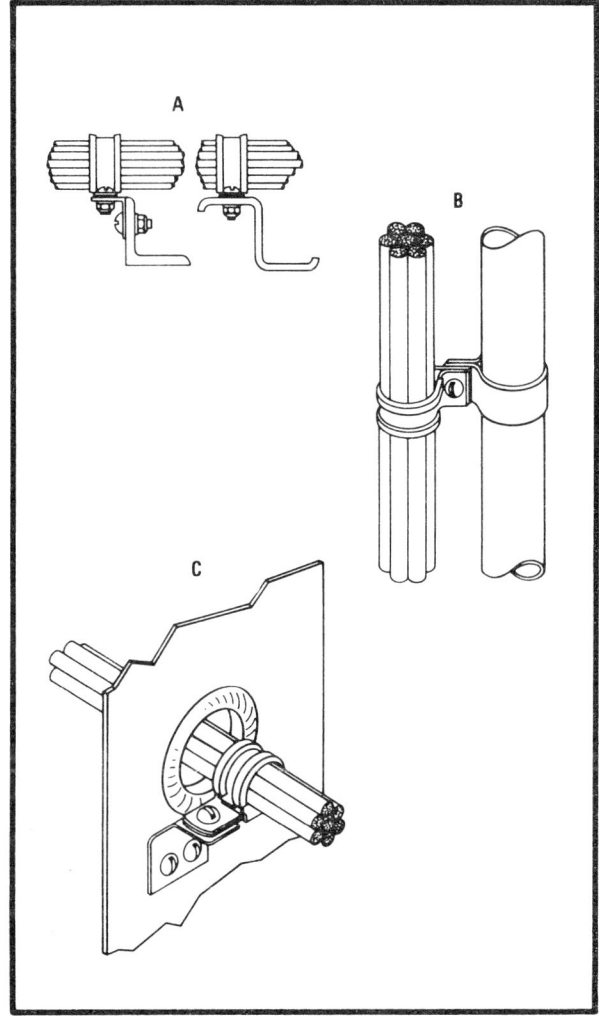

Fig. 116 Bundle attachment to structure.
A - Attachment to sheet metal structural members.
B - Attachment to tubular structure.
C - Support of bundle passing through a hole in a bulkhead or frame.

When there is a long run of bundled wire, it must have no more than a half-inch of slack between adjacent supports; and it if becomes necessary to splice individual wires in a bundle, the splices should be staggered so that there will be no large knot in the bundle, Fig. 117.
Be sure that the wire bundles do not pass through areas where the temperature is high, and if they must be routed parallel with fluid lines, be sure

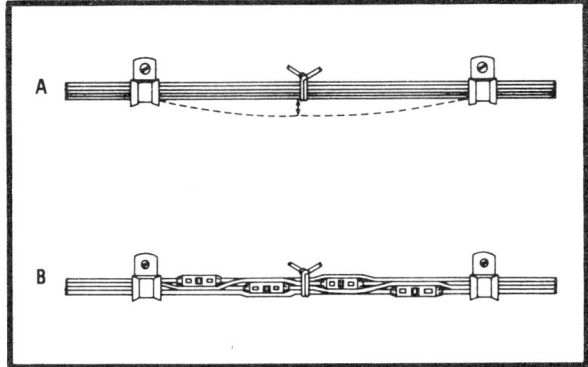

Fig. 117 Wire bundle installation.
A - Bundles should be tight enough that normal hand pressure will deflect them only about a half inch.
B - If splices are made in a bundle, they should be staggered so they will not make a large knot.

the wire bundle is above the line and never supported from it.

B. MOUNTING

Most electronic installations will be made according to drawings approved for the particular installation. These installations must be made in accordance with all of the good structural considerations required for any kind of airframe alteration or repair. The standard criterion for strength of installation is that the mounting should support the following loads:

Forward - 2.0 G (twice the weight of the installed equipment)
Sideways - 1.5 G
Upward - 3.0 G
Downward - 6.6 G

If, for example the installed equipment weighs 12 pounds, the mounting must support 79.2 pounds downward, 36 pounds upward, 24 pounds forward, and 18 pounds sideways.

Any shock mounts used to support the equipment must be of a size that will carry the weight of the installation and installed with the proper orientation of the load. Machine screws and nut plates should be used instead of sheet metal screws.

Cooling is a major consideration with any electrical equipment. Tube-type equipment generates much heat, as the tubes must all have a red-hot filament, and while transistors (better known as solid state equipment) do not generate as much heat, what heat is generated must be carried away, since solid state equipment is far less tolerant of heat than tube type.

Most manufacturers' installation kits include a cooling package. This is usually an air scoop to fit on the outside of the airplane, with the appropriate hoses to bring ram air into a cooling panel which fastens to the side of the radio rack. In flight, air is scooped up and directed through the electronic equipment. When installing this type of cooling kit, be sure the scoop is installed in such a way that any water scooped up will drain outside the airplane, rather than being allowed to get into the equipment.

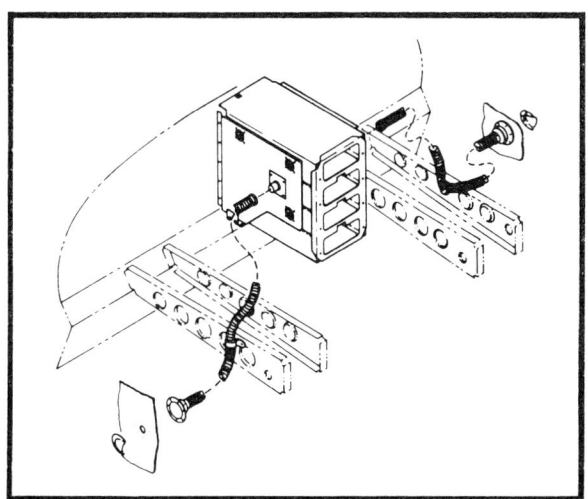

Fig. 118 Heat is the enemy of all electronic equipment and provisions such as this cooling kit installation must be made to remove heat from the equipment.

C. ANTENNA INSTALLATION

An aircraft antenna must not only be capable of radiating or receiving the electronic signal, but it must be structurally strong enough to withstand all of the airloads placed on it in flight and capable of carrying any ice load that is apt to be imposed on it. Usually an antenna mounted on the skin of an airplane will require a doubler inside the structure to strengthen the skin and spread out any stresses imposed on it by the antenna.

Special care must be taken to insure that the antenna is installed in a location approved for the particular airplane. Some locations may cause the antenna to be blanketed by the airplane in certain flight attitudes, and this should be avoided. Be

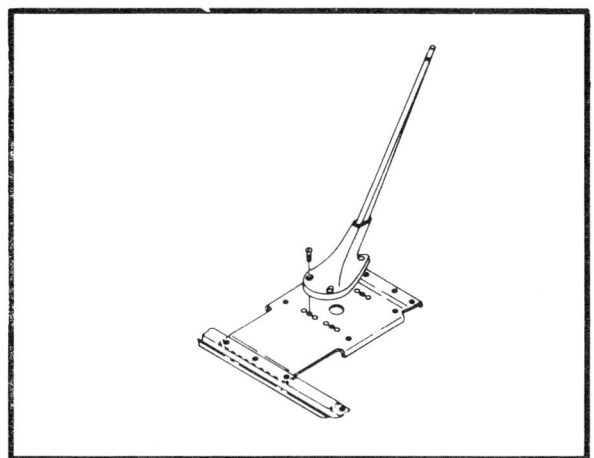

Fig. 119 When an antenna is mounted on an airplane, there should be a doubler installed inside the structure to prevent the skin cracking.

sure that there is no change in reception whether the landing gear is up or down.

D. ELECTRICAL AND MAGNETIC INTERFERENCE

Before any installation may be considered complete, a careful check must be made to be sure that there is no interference between the aircraft or engine and the electronic equipment, and that there is no interference between pieces of installed equipment, themselves.

Alternators or generators are a logical source of electrical interference. Rotating beacons, strobe lights, dynamotors, inverters, electric fuel pumps, flap and gear motors—all are likely to cause interference with radio reception. The manufacturer usually has filters available for offending motors, etc., but if these filters do not eliminate the static, there is a possibility that the trouble is with worn brushes or some other impending malfunction in the motor.

The ignition system is a potent radio transmitter, but almost all aircraft ignition systems are shielded; this means that all of the spark plugs, spark plug leads, magnetos, and switch leads are enclosed in a metal shielding and the interference from the spark is passed off to the ground. Here again, the manufacturer has filters for these sytems that should eliminate the interference if the ignition system proves to be its source. One word of caution: do not use any filter for an ignition system, a generator, or alternator that has not been approved by the manufacturer and the FAA.

After the installation of any electrical or electronic equipment, before the airplane is approved for return to service, be sure to check the magnetic compass. All wires that carry electricity have a magnetic field surrounding them and, while weak, it is enough to deflect the magnetic compass. If the compass indication is different from that which appears on the compass correction card, the compass must be recalibrated, and a new correction card made and installed.

E. PAPERWORK

It is axiomatic that no job is completed until all of the paperwork is done. The installation of any electronic equipment constitues an alteration to the airframe, and, as such, must be recorded. Now, if this installation has been included on the equipment list of the airplane, it is considered to be a minor alteration, and an entry in the maintenance records updating the equipment list, and any appropriate change made in the weight and balance record are all that is required. If, however, the installation is not on the equipment list, the alteration is major, and approval by the FAA is required.

If the installation has been done according to approved data, such as a Supplemental Type Certificate, or a manufacturer's kit, it may be approved by an Authorized Inspector; but if you have no approved data, it must be give a field approval by the FAA district safety agent. In either event an FAA Form 337 (Major Repair and Alteration form) must be completed in duplicate, the original copy staying with the aircraft records, and a copy mailed to the FAA district office. (It is always a good idea to make a third copy to keep for your own personal record.) The weight and balance record must be updated and any required

placards or limitations installed on the instrument panel where the pilot can see them.

While not required by any government regulation it is only wise and professional to furnish the customer with a good, complete wiring diagram, and a bill of materials with any installation so that any future maintenance may be quick and easy, rather that a guessing game.

QUESTIONS

76. What can be done to prevent a ground loop in the shielding when installing a piece of shielded wire?

77. What type of bonding braid is used to bond an aluminum alloy component to an aluminum alloy structure?

78. What type of knot is used when making individual string ties in a wiring bundle?

79. If a piece of electronic equipment weighs 20 pounds, what must be the proven strength of the installation against downward loads?

80. Where does most of the cooling air come from for cooling electronic equipment on light general aviation aircraft?

81. What is usually installed to prevent brush arcing from an electric motor interfering with an electronic installation?

82. What FAA form must be completed before an electronic installation, not included in the aircraft equipment list, can be approved?

Glossary

This glossary of terms is provided to serve as a ready reference for the words with which you may not be familiar. These definitions may differ from those of standard dictionaries, but are in keeping with shop usage.

ADF Automatic direction finder. A method of radio navigation that keeps the pilot informed of his heading relative to the station.

ambiguity, 180° An error inherent in radio direction finding systems, in which the system is unable to determine whether the station is ahead of the airplane or behind it.

amplifier A device or circuit which produces an enlarged version of the input.

antenna, blade A wide-band quarter-wave length antenna.

antenna, current-fed A half-wave antenna, fed in its center.

antenna, dipole A center-fed half-wave antenna.

antenna, goniometer A fixed loop used by automatic direction finding equipment, consisting of two coils, oriented 90° to each other.

antenna, Hertz A half-wave antenna.

antenna, loop A highly direction-sensitive antenna, wound in a coil form.

antenna, Marconi A quarter-wave antenna, utilizing a ground plane, which serves as a quarter-wave reflector.

antenna, stub A short UHF quarter-wavelength antenna, normally used for radar beacon transponders or distance measuring equipment.

antenna, voltage-fed An antenna fed at its end, where the voltage is highest.

antenna, whip A quarter-wave antenna, usually in the high- or very-high-frequency range. It is normally vertically polarized.

area, depletion That area on both sides of the junction of a semiconductor, which varies its characteristics between acting as a conductor and an insulator.

arsenic An element having five valence electrons. When used to dope silicon or germanium, it produces an N-type material.

astable Having no stable conditions. An astable multivibrator is a free-running multivibrator.

azimuth Angular measurement in a horizontal plane.

balun A type of transformer used to match a **balanced** antenna to an **unbalanced** transmission line.

base The center electrode of a transistor. The signal is normally applied to the base.

beacon, marker A highly-directional 75-megahertz signal, used to pinpoint location along an instrument landing system approach.

bias An electrical reference used to establish the operating condition of a semiconductor device or an electron tube.

bias, forward The polarity relationship between a power supply and a semiconductor that allows conduction.

bias, reverse The polarity relationship between a power supply and a semiconductor that does not allow conduction.

bistable A condition that exists in a circuit in which either of two conditions may exist as a steady state.

boron An element having three valence electrons. When it is used to dope silicon or germanium, it produces a P-type material.

cable, coaxial A transmission line in which the center conductor is surrounded by an insulator and a braided outer conductor. All of this is enclosed in a weatherproof outer insulator.

capacitance That ability of an insulator to store electrical energy in the form of electrostatic fields.

capacitor A device used to store electrical energy in the form of electrostatic fields. A capacitor is essentially two conductors separated by an insulator.

cathode The negative terminal of a semiconductor diode, or the element in an electron tube, from which the electrons are emitted.

CDI Course deviation indicator. The instrument used for flying along a VOR. Also called a left-right indicator.

code, Morse A system of dots and dashes, used in aviation to identify ground radio facilities.

collector The electrode in a transistor through which conventional current leaves the transistor.

crystal 1. A thin piece of piezoelectric material having a specific resonant frequency used to control the frequency of an oscillator.
2. A small piece of galena, or lead sulfide, which will allow electron flow in one direction only.

diode A two-element electron tube, or a simple semiconductor device, which allows electron flow in one direction only.

diode, light emitting A semiconductor diode which emits light when current flows through it.

diode, tunnel A special form of semiconductor diode that exhibits a negative resistance characteristic. Under certain conditions, an increase in voltage across the tunnel diode results in a decrease in current through it.

diode, zener A semiconductor diode having a specific peak inverse voltage. It acts as a check valve until this voltage is reached, and then it breaks down and allows flow.

DME Distance measuring equipment. A portion of the military TACAN navigation system used by civilian aircraft to determine the distance in nautical miles from the station.

discriminator A circuit in an FM receiver that converts frequency deviations into amplitude deviations.

doubler, voltage A circuit which produces an output voltage twice that of the input.

drain The electrode in a field effect transistor (FET) that corresponds to the collector of the ordinary transistor.

electron The elementary particle of negative electricity. The electron is that which actually flows in a circuit.

emitter The electrode of a transistor that corresponds to the cathode of a vacuum tube.

filament The small-resistance wire heater inside a vacuum tube, used to heat the cathode so it will emit electrons.

filter, capacitor-input A network consisting of a capacitor and inductor, used to smooth the ripple output of a rectifier.

filter, L An inductor-input filter consisting of an induction and a capacitor used to smooth the ripple from the output of a rectifier.

filter, pi A network consisting of two capacitors and an inductor arranged in the form of the greek letter pi. It is essentially a capacitor-input filter, followed by an L-filter.

frequency, audio Frequency to which the human ear can respond. Essentially from about 16 cycles per second to 16,000 cycles per second.

frequency, intermediate A frequency generated in a superheterodyne receiver equal to the difference between the received radio frequency signal and that produced by the local oscillator.

frequency, radio That frequency of alternating current which causes radio waves to propagate from a conductor.

frequency, resonant The frequency to which a device is tuned. At resonance, the inductive and capacitive reactances are equal.

gate 1. A logic device having one or more inputs and one output. The condition of the inputs determines whether or not a voltage is present at the output.

2. The electrode of a silicon-controlled rectifier or a triac through which the trigger pulse is applied.

gate, AND A logic device which requires a voltage on all of the inputs in order to have a voltage at its output.

gate, EXCLUSIVE OR A logic device having two inputs on which a voltage on either input— but not both—will produce a voltage at the output.

gate, NAND A NOT AND logic device, which will have a voltage at its output only until a voltage appears at all of its inputs.

gate, NOR A NOT OR logic device which will have a voltage at its output only when there is no voltage on any input.

gate, NOT A logic device having one input and one output. There will be no voltage on the output, when a voltage appears at the input.

gate, OR A logic device that will have a voltage on its ouput anytime a voltage appears at any one of its inputs.

germanium An insulating element having four valence electrons used to manufacture semiconductor devices.

glide slope That part of an instrument landing system which provides the pilot with a radio beam to follow in his descent from the outer marker to the point of touchdown.

grid, control The electrode in a vacuum tube to which the signal is applied.

grid, screen The electrode in a tetrode vacuum tube, used to minimize the interelectrode capacitance between the plate and the control grid.

grid, suppressor The electrode in a pentode vacuum tube used to suppress secondary emissions from the plate.

ground plane The reflector used in a quarter-wave antenna, which serves as an additional quarter-wave element.

hertz—Hz A measure of frequency equal to one cycle per second.

heterodyne To mix, or beat together, two frequencies to produce an intermediate frequency.

hill, potential The apparent voltage produced in a semiconductor device by the depletion area. Before electrons can flow in the device, there must be enough voltage applied to overcome the potential hill.

hole A mobile vacancy in the valence structure of a semiconductor that acts as a positive charge.

ILS Instrument landing system. An electronic system used to provide direction and descent guidance for instrument landings. An ILS consists of localizer, glide slope, marker beacons, and approach lights.

impedance The total opposition to the flow of alternating current. Impedance is the vector sum of reactance and resistance.

inductance Opposition to the flow of alternating current or changing direct current caused by the magnetic fields surrounding the conductor.

inductor A coil or other device used to introduce inductance into a circuit.

kilohertz—kHz 1,000 cycles per second.

limiter A stage in a frequency modulated receiver that limits the amplitude of the signal and thus removes static.

line, transmission A conductor, usually coaxial, used to join a receiver or transmitter to the antenna.

locator, compass A low-frequency nondirectional beacon, co-located with the marker beacons used to help establish the pilot on the localizer for an ILS approach.

localizer That portion of an instrument landing system used to give the pilot directional guidance down the center-line of the instrument runway.

logic The science of dealing with the basic elements of truth and the use of truth tables.

megahertz—MHz 1,000,000 cycles per second.

modulation The changing of frequency or amplitude by superimposing an audio frequency on the carrier frequency.

modulation, amplitude—AM The system of radio information transmissions in which the amplitude of the carrier is changed by the radio frequency.

modulation, frequency—FM The method by which information is transmitted by varying the frequency of the carrier with the audio.

monostable The condition of a device which has one stable condition. When disturbed from this, it will return to its original condition.

multivibrator A form of oscillator which produces its output by having one or another transistor conduct. When one conducts, the other is shut off. Conduction alternates between the two.

OBS Omni bearing selector. The variable phase shifter in a VOR system with which the pilot selects the radial on which he wishes to fly.

oscillator An electronic device which converts direct current into alternating current.

oscillator, Hartley The form of electronic oscillator which produces its feedback through a tapped inductor.

oscillator, multivibrator A form or electronic oscillator in which two semiconductors alternately conduct.

oscillator, relaxation A form of electronic oscillator which produces a sawtooth waveform by charging a capacitor, then rapidly discharging it.

pentode A five-element vacuum tube.

photodiode A semiconductor diode which can be caused to conduct in its reverse direction by the application of light to its junction.

phototransistor A transistor which can be for-biased into conduction by the application of light to its emitter-base junction.

plate The electrode in a vacuum tube which receives electrons from the cathode.

ratio, voltage standing wave—VSWR The ratio of the maximum voltage to the minimum voltage along a coaxial cable.

reactance, capacitive The opposition to the flow of alternating current electricity caused by the capacitance in a circuit.

reactance, inductive The opposition to the flow of alternating current caused by the inductance in a circuit.

rectifier, bridge A form of rectifier using four diodes arranged in a bridge circuit.

rectifier, full-wave A form of rectifier which inverts one-half of the input AC signal and provides a pulsating DC output having twice the frequency of the input AC.

rectifier, half-wave A form of rectifier using one diode that produces only one-half of the alternating current wave in its output.

rectifier, silicon-controlled A form of gated rectifier that allows current to flow only during that portion of the cycle after which the gate has been triggered by a positive pulse.

resonance A circuit condition in which inductive and capacitive reactances are equal.

semiconductor A solid state device that will conduct when forward-biased and will not conduct when reverse-biased.

silicon A natural element having four electrons in its valence orbit. Silicon is used to produce semiconductor devices having excellent thermal characteristics.

silicon, N-type Silicon which has been doped with a pentavalent impurity.

silicon, P-type Silicon which has been doped with a trivalent impurity.

source The electrode of a field-effect transistor which compares to the emitter of an ordinary transistor.

squelch A circuit in a communications receiver which holds the output volume down until a signal is received.

superheterodyne A receiver circuit in which the received radio frequency signal is mixed with a frequency produced in a local oscillator to create an intermediate frequency. This **IF** is amplified, detected and demodulated to produce the modulating audio frequency in its output.

tetrode A four-element vacuum tube.

transistor, field effect A special form of semi-conductor device with a high input impedance. Electron flow between its source and drain is controlled by a voltage applied to the gate.

transistor, NPN A three-element semiconductor made up of a sandwich of P-type silicon or germanium between two pieces of N-type material.

transistor, PNP A three-element semiconductor device made up of a sandwich of N-type silicon or gernamium between two pieces of P-type material.

transistor, unijunction A special form of transistor which allows flow between its two bases when an appropriate voltage is applied to its emitter.

triac A semiconductor device similar to an SCR that may be triggered by either a positive or a negative pulse applied to its gate.

triode A vacuum tube having three active electrodes.

tube, vacuum An evacuated glass or metal envelope containing a cathode, a heater, a plate, and often one or more grids. It serves as an electron control valve.

voltage, avalanche The reverse voltage required to cause a zener diode to break down.

VOR Very high frequency omnirange navigation system. A phase comparison system used to provide directional guidance for aircraft.

wave, carrier High frequency alternating current which can be modulated to carry intelligence by propagation as a radio wave.

wave, electric One of the components of a radio wave produced along the length of the antenna.

wave, magnetic One of the components of a radio wave that is perpendicular to the antenna.

wave, sawtooth The waveform produced by the relaxation oscillator in which the voltage rises slowly and drops off rapidly.

wave, sine The waveform produced by a rotary generator in which the amplitude of the wave varies as the sine of the angle through which the generator has turned.

wave, square The waveform of a multivibrator oscillator in which the leading edge and trailing edge of the wave are both vertical.

Answers To Study Questions

1. Current changes lag voltage changes in an inductive circuit.

2. Smaller.

3. Impedance.

4. Resonant frequency.

5. High.

6. It minimizes the feedback problem caused by interelectrode capacitance between the plate and control grid.

7. Prevents, or minimizes, secondary emission from the plate which would cause excessive screen grid current.

8. N-type silicon has an excess of electrons.

9. The trivalent impurity does not have enough valence electrons to form covalent bonds with all of the silicon, and a space is left in which electrons can go.

10. Reverse.

11. Forward.

12. The polarities are reversed between the PNP and the NPN.

13. SCR's can be used to dim lights by varying the portion of the cycle in which current is allowed to flow.

14. Positive.

15. A triac may be caused to conduct with either a positive or a negative pulse, and thus controls both halves of the cycle.

16. Conduction through an FET is controlled by the voltage on its gate.

17. In a certain portion of the operating characteristics of a tunnel diode, an increase in voltage will cause a decrease in current.

18. A UJT will conduct when the voltage at its emitter becomes sufficiently high to trigger it.

19. A photodiode is caused to conduct when sufficient light energy strikes its junction.

20. A phototransistor will conduct much more current than a photodiode.

21. It is forward-biased so current can flow through it, causing it to emit photons.

22. Half-wave rectifiers use only one-half of the AC cycle.

23. All of the AC is used, and the resulting higher frequency pulsating DC is easier to filter than that produced by a half-wave rectifier.

24. It converts all of the secondary voltage into DC, rather than just half of it through the center-tapped transformer.

25. Six.

26. Yes.

27. High.

28. No.

29. When the neon bulb conducts, it short-circuits the charged capacitor.

30. a. There must be amplification.
    b. There must be feedback of the proper phase from the output, back into the input.

31. The inductance and capacitance in the tank circuit.

32. Astable means free running. It will continue to oscillate as long as power is supplied.

33. Either transistor will conduct, but not both at the same time.

34. One transistor will conduct until a pulse shuts it off, and causes the other to conduct. After a specific time period, the second transistor will shut off, and the first conduct again.

35. A Zener diode.

36. 26

37. 01100

38. Yes.

39. VHF

40. The amplitude, or the voltage of the signal.

41. When two frequencies are fed into a mixer, they will beat together, or heterodyne, and produce two more frequencies, the sum of the two, and their difference.

42. A constant frequency, which is the difference between the frequency of the received RF and that of a local oscillator.

43. Automatic Volume Control adjusts the gain in a receiver to keep the output constant as the strength of the input signal varies.

44. Double conversion allows more precise control of frequency and makes possible more channels in the same frequency allocation.

45. The frequency of the carrier varies with the amplitude of the signal.

46. The oscillator rather than the power amplifier is modulated.

47. a. There must be an oscillator to generate the appropriate carrier frequency.
    b. There must be amplifiers.
    c. There must be a modulation system to modulate or change the carrier.
    d. There must be an antenna to radiate the energy.

48. VHF is line of sight communications.

49. One-half wave length.

50. 6.24 feet, or 74.88 inches.

51. The ground plane serves as a reflector, allowing a quarter-wave-length antenna to be used.

52. When it is pointed toward the station.

53. Coaxial cable has no external field, and it is not susceptible to external fields.

54. A balun connects a balanced antenna to an unbalanced transmission line.

55. A low frequency receiver and a good ear.

56. A radio direction finder cannot determine whether the station is ahead or behind the airplane. This is called 180° ambiguity.

57. A horizontally polarized VHF dipole with the rods arranged in a V.

58. A direction sensitive loop, and an omnidirectional sense antenna.

59. The cardiod shaped field pattern of the combined loop and sense antennas.

60. a. Localizer
    b. Glide slope
    c. Marker beacons

61. The localizer is four times as sensitive as the VOR.

62. The ADF receiver.

63. 75 megahertz.

64. A UHF dipole.

65. It is paired with the localizer and automatically tuned when the localizer frequency is selected.

66. A short, vertically polarized UHF stub, or blade.

67. The light blinks each time the transponder replies to an interrogation from the ground.

68. He momentarily depresses the Ident button, when requested to do so.

69. An encoding altimeter.

70. The same type of vertically polarized UHF stub or blade antenna used by the DME.

71. 10 ampere.

72. 7.5 amperes.

73. 0 gage.

74. 10 gage.

75. Blue.

76. Ground the shielding at one end only.

77. Aluminum alloy.

78. A clove hitch secured with a square knot.

79. 132 pounds.

80. Ram air picked up in scoops on the outside of the airplane.

81. A filter, usually a capacitor-inductor type.

82. FAA form 337. ''Major Alteration and Repair'' form.

Basic Electronics and Radio Installation Final Examination

STUDENT_____

GRADE _____

Place a circle around the letter for the answer which is most nearly correct.

1. What type of device is used in a transistor voltage regulator to sense the voltage?

   A. Silicon-controlled rectifier
   B. Field effect transistor
   C. Zener Diode
   D. N-P-N transistor

2. What is the basic difference between a triac and a silicon-controlled rectifier?

   A. A triac requires a negative pulse to trigger it, while an SCR uses a positive pulse.
   B. A triac may be triggered with a pulse of either polarity, while an SCR requires a positive pulse.
   C. SCR's are used in AC circuits, while triacs may be used in either AC or DC circuits.
   D. Triacs are used for only small currents, while SCR's are used for heavy current applications.

3. Which device has the lowest input impedance?

   A. N-P-N transistor
   B. Field effect transistor
   C. Triode vacuum tube
   D. Reverse-biased diode

4. What type of device may be used to provide a pulse of energy to trigger the gate of a silicon-controlled rectifier?

   A. Field effect transistor
   B. Phototransistor
   C. Zener diode
   D. Unijunction transistor

5. How many semiconductor diodes are required for a three-phase full-wave rectifier?

   A. Four
   B. Six
   C. Three
   D. Twelve

6. What shape waveform is produced by a multivibrator oscillator?

   A. Sine wave
   B. Sawtooth wave
   C. Square wave
   D. Pure DC

7. What is the binary equivalent of 56?

   A. 111000
   B. 010100
   C. 110100
   D. 001110

8. What kind of logic gate produces a voltage at its output when there is a voltage on one, and only one, of its inputs?

   A. AND gate
   B. NOR gate
   C. OR gate
   D. EXCLUSIVE OR gate

9. What is the purpose of a screen grid in a vacuum tube?

   A. It controls the flow of electrons between the cathode and the plate.
   B. It minimizes the feedback through inter-electrode capacitance.
   C. It suppresses secondary emission from the plate.
   D. It provides electrons for secondary emission.

10. Which type of transistor amplifier inverts the phase of the amplified signal?

    A. Common collector
    B. Common base
    C. Common emitter
    D. All of the above.

11. Which equipment would require the longest antenna?

    A. Automatic direction finder (ADF)
    B. Distance measuring equipment (DME)
    C. Radar beacon transponder
    D. ILS localizer

12. Which two pieces of equipment use the same antenna?

    A. VOR and DME
    B. Localizer and transponder
    C. ADF and VOR
    D. VOR and localizer

13. Which two forms of radio navigation compare voltages?

    A. VOR and ADF
    B. Localizer and glide slope
    C. DME and transponder
    D. Marker beacon and compass locators.

14. What kind of antenna is used with VOR?

    A. Long wire sense Antenna and a loop
    B. Horizontally polarized V-dipole
    C. Vertically polarized VHF whip antenna
    D. Servo-driven loop

15. How can a ground loop in a shielded wire installation be avoided?

    A. Use shielded wire only for carrying DC.
    B. Ground the shield at both ends.
    C. Ground the shield at only one end.
    D. Do not ground the shield.

16. Using the wire chart, find the appropriate size wire to use in a 28-volt system for carrying 15 amperes continuous in a bundle for 50 feet.

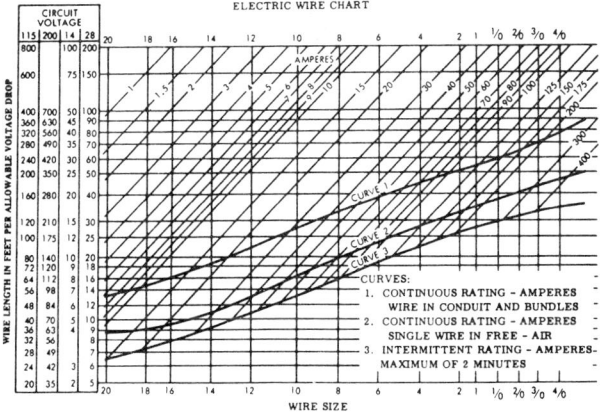

A. 12-gage
B. 8-gage
C. 14-gage
D. 10-gage

17. What is the allowable voltage drop in the wiring for an intermittent load in a 115-volt electrical system?

    A. 8 volts
    B. 4 volts
    C. 6 volts
    D. 2 volts

18. What is the color code for a pre-insulated terminal to fit a 20-gage wire?

    A. Blue
    B. Red
    C. Yellow
    D. Green

19. What device is used with a loop antenna to prevent 180° ambiguity?

    A. Goniometer
    B. Preselector
    C. Interrogator
    D. Sense antenna

20. What type of antenna is used with distance measuring equipment?

    A. Vertically polarized UHF blade
    B. Horizontally polarized 75-MHz fixed frequency antenna
    C. Vertically polarized VHF blade
    D. Horizontally polarized UHF dipole

NOTES

Basic Electronics and Radio Installation Answers To Final Examination

1. C

2. B

3. A

4. D

5. B

6. C

7. A

8. D

9. B

10. C

11. A

12. D

13. B

14. B

15. C

16. D

17. A

18. B

19. D

20. A

Appendix: Electronic Symbols

A. GENERAL SYMBOLS

1. Resistors

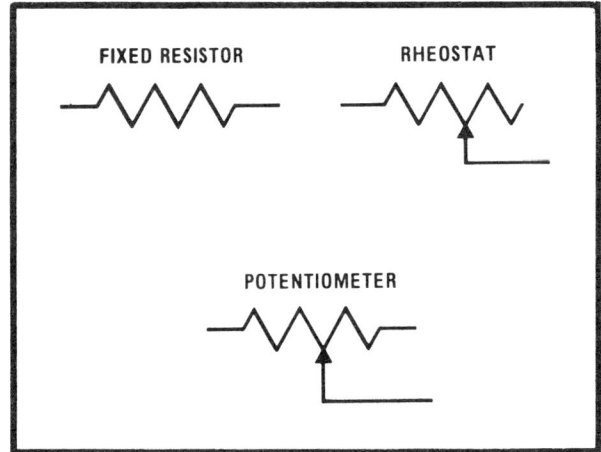

2. Wire

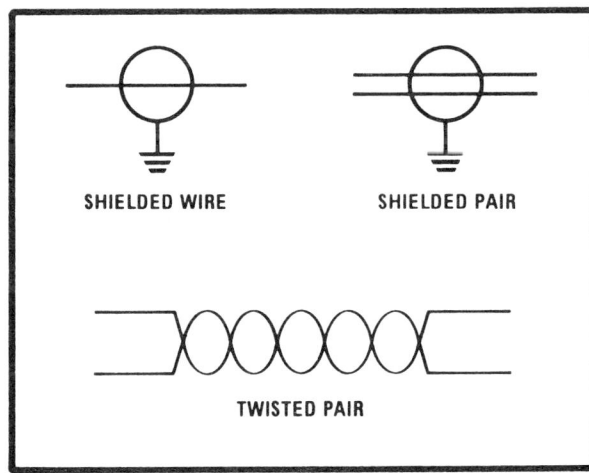

3. Capacitors

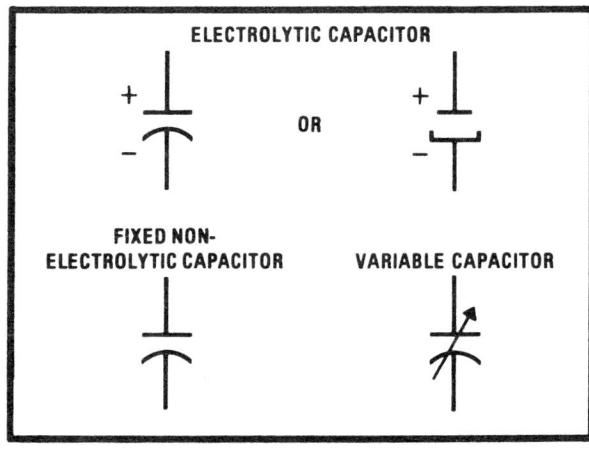

4. Inductors

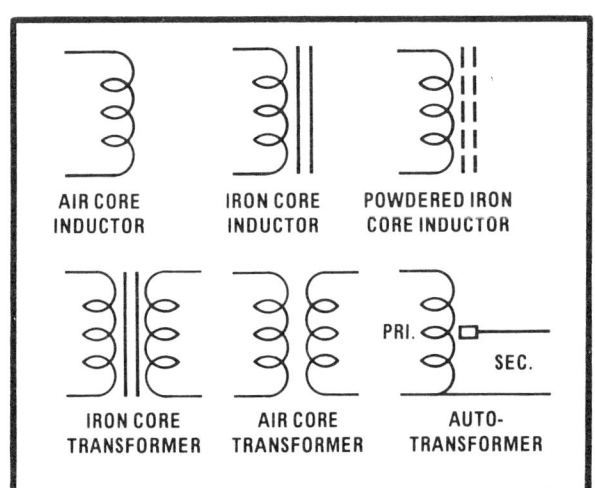

5. Crystal

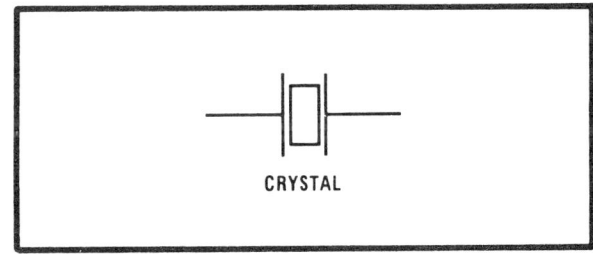

6. Fuse

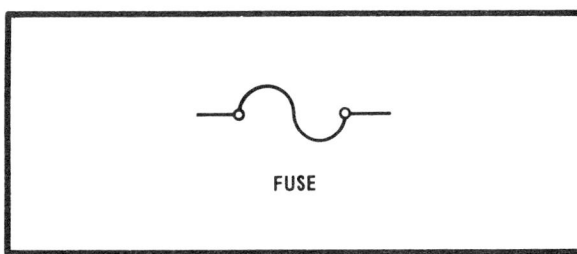

7. Grounds

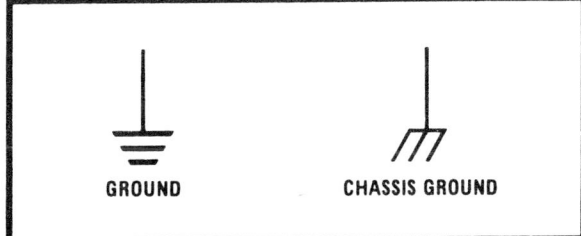

B. SEMICONDUCTOR DEVICES

1. Diodes

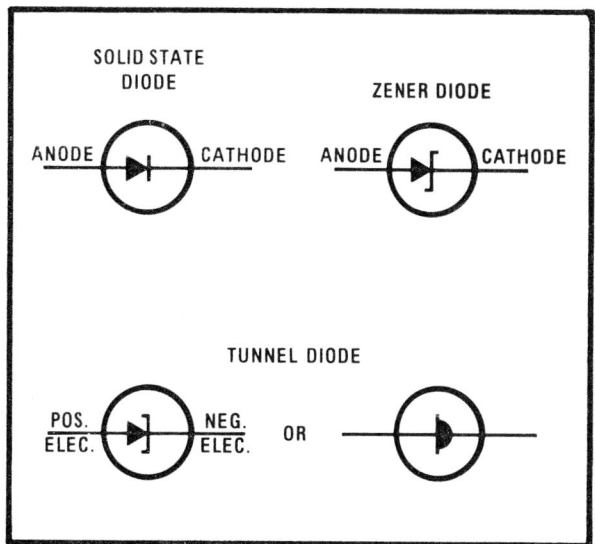

2. Transistors

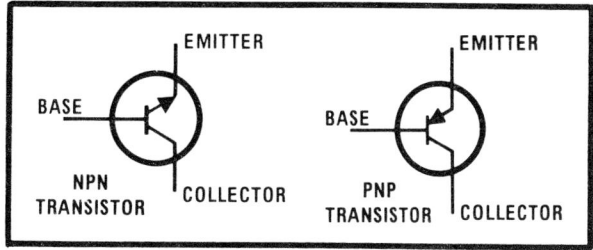

3. Field-Effect Transistors

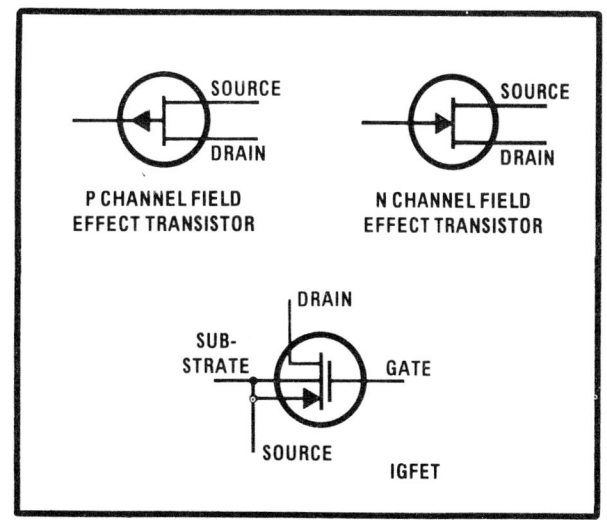

4. Gated Devices

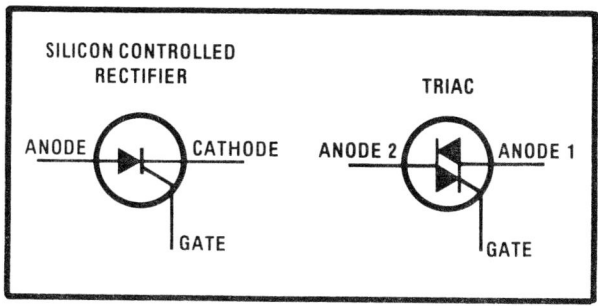

5. Thyrector

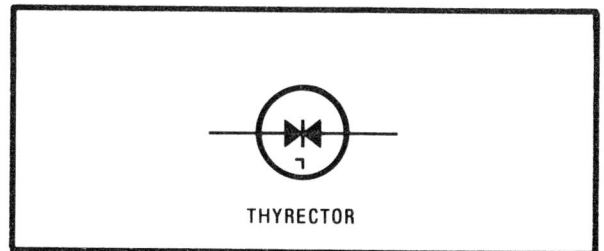

6. Unijuction Transistor

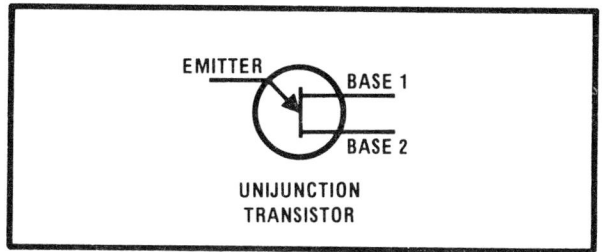

7. Optoelectronic Devices

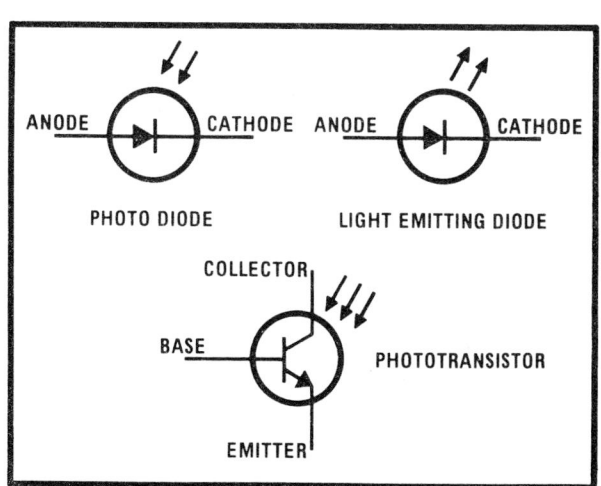

C. VACUUM TUBES

NOTES:

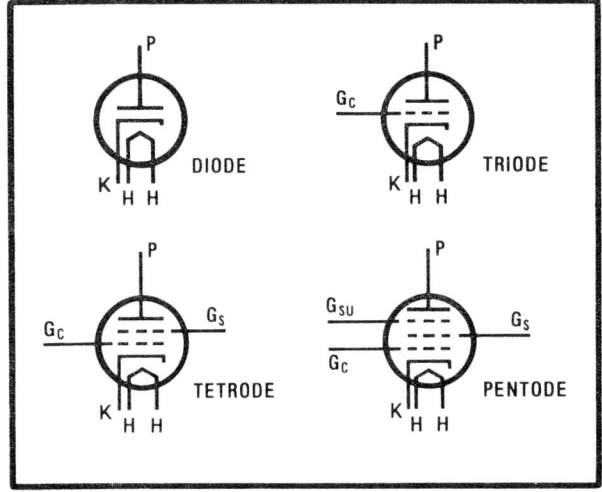

NOTES: